Nonlinear Optics - Nonlinear Nanophotonics and Novel Materials for Nonlinear Optics

Edited by Boris I. Lembrikov

Published in London, United Kingdom

IntechOpen

Supporting open minds since 2005

Nonlinear Optics - Nonlinear Nanophotonics and Novel Materials for Nonlinear Optics
http://dx.doi.org/10.5772/intechopen.94637
Edited by Boris I. Lembrikov

Contributors
Helena Cristina Vasconcelos, Mohsen Lutephy, Ana-Maria Albu, Vlad Marian Târpă, Emusani Ramya, Muhammad Aamir Iqbal, Naila Ashraf, Wajeehah Shahid, Abdullah Khan Durrani, Khurram Shahzad, Mujtaba Ikram, Muhammad Awais, Boris I. Lembrikov

First published in London, United Kingdom, 2022 by IntechOpen
IntechOpen is the global imprint of INTECHOPEN LIMITED, registered in England and Wales, registration number: 11086078, 5 Princes Gate Court, London, SW7 2QJ, United Kingdom
Printed in Croatia

British Library Cataloguing-in-Publication Data
A catalogue record for this book is available from the British Library

Additional hard and PDF copies can be obtained from orders@intechopen.com

Nonlinear Optics - Nonlinear Nanophotonics and Novel Materials for Nonlinear Optics
Edited by Boris I. Lembrikov
p. cm.
Print ISBN 978-1-83962-835-1
Online ISBN 978-1-83962-836-8
eBook (PDF) ISBN 978-1-83962-890-0

For EU product safety concerns:
IN TECH d.o.o., Prolaz Marije Krucifikse Kozulić 3, 51000 Rijeka, Croatia, info@intechopen.com or visit our website at intechopen.com.

Meet the editor

Dr. Boris I. Lembrikov is a senior lecturer at the Holon Institute of Technology (HIT), Israel. He is a co-author of *Electrodynamics of Magneto-active Media* and author of forty-five papers published in peer-reviewed international scientific journals, sixteen chapters in scientific books, and thirty-eight papers in conference proceedings. He has also edited three books. Dr. Lembrikov's research interests include nonlinear optics, optical communications, quantum dot lasers, optical amplifiers, and plasmonics.

Contents

Preface

Nanophotonics has recently attracted interest for its potential applications in microscopy, sensing, imaging, medicine, light sources, and functional devices. These applications are based on the interaction of light and matter at the nanoscale. The plasmonic oscillations in metallic structures are essential in such a case due to the strong enhancement of the light-matter interactions near the metal surface in small volumes of nanostructures. Plasmonic excitations at the interface of the metal and dielectric or surface plasmon polaritons (SPPs) can exist as transverse magnetic (TM) waves. SPPs are characterized by limited propagation length and confinement at the subwavelength scale in the direction perpendicular to the propagation direction. The nonlinear optical effects are enhanced in plasmonic structures due to the strong localized electromagnetic fields and the influence of the surrounding medium on the plasma excitation parameters. Plasmonic applications in nonlinear optics result in the enhancement of optical nonlinearity, substantial decrease of photonic device dimensions, and the possibility of ultrafast optical signal processing.

This book examines nonlinear optical effects in nonlinear nanophotonics, plasmonics, and novel materials for nonlinear optics. It includes six chapters organized into four sections.

Section 1 includes an introductory chapter that investigates the interaction of an electromagnetic field and free electron gas in metals. The analysis is based on the solution of Maxwell's equations and the equation of motion for a free electron in an external electric field. As a result, different types of plasmonic excitations such as volume plasmons, localized surface plasmons, and surface plasmon polaritons emerge. The introductory chapter also discusses the peculiarities of the nonlinear optical phenomena in plasmonic nanostructures and metamaterials.

Section 2 includes Chapter 2, which considers the foundations, the peculiarities of nonlinear optical phenomena in nanophotonics, and the possible applications of nanophotonics.

Section 3 consists of three chapters that discuss novel materials for nonlinear optics. Chapter 3 presents a new method of preparing metal nanoparticles and investigates the nonlinear properties of these nanoparticles experimentally. Chapter 4 studies the structural-compositional relationship of polymeric materials used in optoelectronic and photonic applications. Chapter 5 reviews the nonlinear optical properties of photonic glasses. In particular, the chapter focuses on optical glasses doped with metal nanoparticles.

Section 4 consists of Chapter 6, which demonstrates a new solution of the nonlinear Schrödinger equation (NLSE) for 5G broadband millimeter waves.

This book is a useful resource that may be interesting for the researchers and engineers occupied studying and working in nonlinear optics, nanophotonics, and devices for optical signal processing.

Boris I. Lembrikov
Department of Electrical Engineering and Electronics,
Holon Institute of Technology (HIT),
Holon, Israel

Section 1

Introduction

Chapter 1

Introductory Chapter: Nonlinear Optical Phenomena in Plasmonics, Nanophotonics and Metamaterials

Boris I. Lembrikov

1. Introduction

The observation of nonlinear optical effects started with the discovery of the laser in 1960 as a source of the high-intensity coherent radiation [1, 2]. Nonlinear optics is the study of the nonlinear light and matter interaction effects such as the light self-focusing and self-trapping, second harmonic generation (SHG), third-harmonic generation (THG), four-wave mixing (FWM), parametric processes, different types of stimulated light scattering (SLS), soliton generation and propagation [1, 2]. The nonlinear optical phenomena are widely used in modern communication systems for such applications as the generation of ultra-short pulses, all-optical signal processing and ultrafast switching [2, 3]. Recently, new fields of nonlinear optics emerged such as strong-field nano-optics, nonlinear plasmonics, and nonlinear metamaterials [3–8].

In the case of the strong-field nano-optics the optical field interacts with matter at the wavelength or even at the subwavelength scale [5]. Strong electromagnetic fields of a light wave with the wavelength $\lambda \sim 1\mu m$ interacting with electrons are confined to a few nanometers which result in a substantial enhancement of the local electric field [4]. Typically, an increase of the nonlinear optical response at a nanoscale is caused by the plasmonic effects, i.e., the coherent oscillations of the conduction electrons near the surface of noble-metal structures [3–6]. At the extended metal surfaces, the surface plasmon polaritons (SPPs) can occur [3]. SPP is a surface electromagnetic wave propagating at the metal-dielectric interface [3–5]. In the case of the metal nanoparticles there exist localized surface plasmons (LSPs) with resonances depending on the nanoparticle size and shape [3]. Nonlinear optical effects can be significantly enhanced by plasmonic excitations in two ways: (i) the plasmonic structures provide the optical field enhancement near the metal-dielectric interface due to SPPs or LSPs; (ii) the SPP and LSP parameters are very sensitive to refractive indices of the metal and the surrounding dielectric medium [3–6]. Plasmonic excitations are characterized by the timescale of several femtoseconds which makes it possible the ultrafast optical signal processing [3].

Metamaterials are artificial materials with desirable properties [7]. For this reason, the magnitude of the metamaterial optical nonlinearity can be substantially increased [7]. Generally, all fundamental nonlinear optical phenomena such as self-action effects, wavelength conversion, nonlinear surface waves, nonlinear guided waves and solitons are possible in nonlinear metamaterials [7]. The nonlinear metamaterials are closely related to plasmonics, active media, and nonlinearity based on liquid crystals (LCs) [7, 8]. LCs possess simultaneously the properties of a liquid and a crystal [8–10]. Thermotropic LCs self-assemble in a different ordered arrangements of their crystalline axis depending on the temperature [8–10]. The orientational order of LC may be changed by a moderate external electric field [8–10]. For this reason, liquid crystalline media are characterized by strong optical nonlinearity [8–10]. Nematic LCs (NLCs) characterized by the ordering of molecular elongated axes are mainly used in applications [8].

In this chapter, we discuss the fundamentals of plasmonics and the specific features of the nonlinear optical effects in plasmonic nanostructures. The chapter is organized as follows. In Section 2 the interaction of the electromagnetic field with the free electrons in metals is described based on Maxwell's equations and equation of motion for a free electron in the external electric field. The dielectric function of the free electron gas is obtained based on the Drude model. In Section 3 SPPs at the interface of a metal and a dielectric are investigated. In Section 4 the nonlinear phenomena in the plasmonic structures are briefly discussed. The details of this topic may be found in the corresponding references. Conclusions are presented in Section 5.

2. Interaction between electromagnetic field and free electrons in metals

The interaction of electromagnetic fields with metals is described by Maxwell's equations of macroscopic electromagnetism [11]. This approach is valid also for metallic nanostructures characterized by the sizes of several nanometers [11]. The Maxwell's equations have the form [1, 2, 11].

$$\nabla \cdot \vec{D} = \rho_{ext} \tag{1}$$

$$\nabla \cdot \vec{B} = 0 \tag{2}$$

$$\nabla \times \vec{E} = -\frac{\partial \vec{B}}{\partial t} \tag{3}$$

$$\nabla \times \vec{H} = \vec{J}_{ext} + \frac{\partial \vec{D}}{\partial t} \tag{4}$$

Here $\vec{D}, \vec{E}, \vec{H}$ and $\vec{B}$ are the dielectric displacement, the electric field, the magnetic field and the magnetic induction (or magnetic flux density), respectively; ρ_{ext} and $\vec{J}_{ext}$ are external charge and current densities, respectively. The four macroscopic fields are related by the following material Equations [11].

$$\vec{D} = \varepsilon_0 \varepsilon \vec{E} \tag{5}$$

$$\vec{B} = \mu_0 \mu \vec{H} \tag{6}$$

where ε_0 and μ_0 are the electric permittivity and magnetic permeability of vacuum, respectively, ε is the relative permittivity and μ is the relative permeability of the medium, $\mu = 1$ for the nonmagnetic medium. The total charge and current densities ρ_{tot} and $\vec{J}_{tot}$ consist of the external charge and current densities ρ_{ext}, $\vec{J}_{ext}$ and the internal ones $\rho, \vec{J}$. They are given by [11]: $\rho_{tot} = \rho_{ext} + \rho$, $\vec{J}_{tot} = \vec{J}_{ext} + \vec{J}$. The internal current density has the form [11].

$$\vec{J} = \sigma \vec{E} \tag{7}$$

where σ is the conductivity of the medium. The optical response of metals typically depends on frequency and wave vector in the case of a spatial dispersion [11]. In such a case, Eqs. (5) and (7) can be generalized to the following relationships taking into account the non-locality in time and space [11].

$$\vec{D}\left(\vec{r},t\right) = \varepsilon_0 \int dt' d\vec{r}' \varepsilon\left(\vec{r} - \vec{r}', t - t'\right)\vec{E}\left(\vec{r}', t'\right) \tag{8}$$

$$\vec{J}\left(\vec{r},t\right) = \varepsilon_0 \int dt' d\vec{r}' \sigma\left(\vec{r} - \vec{r}', t - t'\right)\vec{E}\left(\vec{r}', t'\right) \tag{9}$$

The Fourier transform of expressions (8) and (9) results in the following relationships in the Fourier domain [11].

$$\vec{D}\left(\vec{k},\omega\right) = \varepsilon_0 \varepsilon\left(\vec{k},\omega\right)\vec{E}\left(\vec{k},\omega\right) \tag{10}$$

$$\vec{J}\left(\vec{k},\omega\right) = \sigma\left(\vec{k},\omega\right)\vec{E}\left(\vec{k},\omega\right) \tag{11}$$

where $\vec{k}$ and ω are the wave vector and angular frequency of the field plane-wave components, respectively.

Combining Eqs. (3) and (4) we obtain the wave equation for the electric field $\vec{E}\left(\vec{r},t\right)$ [1, 2, 11].

$$\nabla \times \nabla \times \vec{E} = -\mu_0 \frac{\partial^2 \vec{D}}{\partial t^2} \tag{12}$$

which takes the form in the frequency domain [11].

$$\vec{k}\left(\vec{k}\cdot\vec{E}\right) - k^2\vec{E} = -\varepsilon\left(\vec{k},\omega\right)\frac{\omega^2}{c^2}\vec{E} \tag{13}$$

where $c = \left(\sqrt{\varepsilon_0 \mu_0}\right)^{-1}$ is the speed of light in vacuum. There are two types of solutions of Eq. (13): the transverse waves where the electric field vector is perpendicular to the wave propagation direction, and the longitudinal waves where the electric field vector is parallel to the wave propagation direction. The transverse waves are characterized by the condition $\left(\vec{k}\cdot\vec{E}\right) = 0$ and the corresponding dispersion relation

$$k^2 = \varepsilon\left(\vec{k},\omega\right)\frac{\omega^2}{c^2} \tag{14}$$

The dispersion relation for the longitudinal waves can be obtained from Eq. (13).

$$\varepsilon\left(\vec{k},\omega\right) = 0 \tag{15}$$

It is seen from Eq. (15) that the longitudinal collective oscillations can exist only for the frequencies ω which correspond to the zeros of the dielectric function $\varepsilon\left(\vec{k},\omega\right)$ [11].

Consider a plasma model that explains the optical properties of alkali metals and noble metals for the frequencies up to the ultraviolet ones and to the visible ones, respectively [11, 12]. In the framework of the phenomenological plasma model the metal crystal lattice potential and electron–electron interactions are not taken into account [11]. It is assumed that the details of the metal energy band structure are included into the effective optical mass m of each electron [11]. In such a model, the equation of motion for an electron with the charge e and the effective mass m of the plasma sea in the external electric field $\vec{E}$ has the form [11].

$$m\frac{d^2\vec{x}}{dt^2} + m\gamma\frac{d\vec{x}}{dt} = -e\vec{E} \tag{16}$$

where $\vec{x}$ is the free electron displacement, $\gamma = \tau^{-1}$ is a characteristic collision frequency, τ is the free electron gas relaxation time. Typically, $\tau \sim 10^{-14} s$ at room temperature which yields $\gamma \approx 100\, THz$ [11]. For the external field $\vec{E}(t) = \vec{E}_0 \exp(-i\omega t)$ we obtain from Eq. (16) the following particular solution.

$$\vec{x}(t) = \vec{x}_0 \exp(-i\omega t) = \frac{e}{m(\omega^2 + i\gamma\omega)}\vec{E}(t) \tag{17}$$

The macroscopic polarization $\vec{P}$ of the medium caused by the displaced electrons is given by [11].

$$\vec{P} = -ne\vec{x} \tag{18}$$

where n is the free electrons concentration. Substituting expression (17) into Eq. (18) we obtain.

$$\vec{P} = -\frac{ne^2}{m(\omega^2 + i\gamma\omega)}\vec{E} \tag{19}$$

Taking into account that the dielectric displacement $\vec{D}$ in a medium is given by [1, 2, 11].

$$\vec{D} = \varepsilon_0\vec{E} + \vec{P} \tag{20}$$

and substituting expression (19) into Eq. (20) we obtain.

$$\vec{D} = \varepsilon_0\left(1 - \frac{\omega_p^2}{\omega^2 + i\gamma\omega}\right)\vec{E} \tag{21}$$

where $\omega_p = \sqrt{ne^2/(\varepsilon_0 m)}$ is the plasma frequency of the free electron gas [4, 11]. Comparing relationships (5), (10) and (21) we obtain the expression of the free electron gas dielectric function $\varepsilon(\omega)$ [4, 11].

$$\varepsilon(\omega) = 1 - \frac{\omega_p^2}{\omega^2 + i\gamma\omega} \tag{22}$$

The dielectric function $\varepsilon(\omega)$ (22) is known as the Drude model of the optical response of metals [11]. Expression (22) can be divided into real and imaginary components as follows.

$$\operatorname{Re}\varepsilon(\omega) = 1 - \frac{\omega_p^2\tau^2}{1+\omega^2\tau^2};\ \operatorname{Im}\varepsilon(\omega) = \frac{\omega_p^2\tau}{\omega(1+\omega^2\tau^2)} \tag{23}$$

Typically, the frequencies $\omega < \omega_p$ are considered where the metals have the pronounced metallic properties since in such a case $\operatorname{Re}\varepsilon(\omega) < 0$ [11]. In the high frequencies limiting case $\omega\tau \gg 1$ the imaginary part of the dielectric function (23) can be neglected while expression (22) takes the form [11].

$$\varepsilon(\omega) \approx 1 - \frac{\omega_p^2}{\omega^2} \tag{24}$$

Comparison of expressions (15) and (24) shows that the longitudinal waves can be excited at the plasma frequency $\omega = \omega_p$. These longitudinal oscillations are called the volume plasma oscillations, and the quasi-particles of these oscillations are called the volume plasmons [11]. The volume plasmons cannot interact with the transverse electromagnetic waves and can be excited by particle impact [11].

3. Surface Plasmon Polaritons (SPPs)

Consider now the SPPs at the interface between the metal $(z < 0)$ and the dielectric $(z > 0)$ shown in **Figure 1**. The metal and the dielectric are characterized by the dielectric function $\varepsilon_1(\omega)$ defined by expression (22) and the real dielectric constant $\varepsilon_2 > 0$, respectively. We consider the frequency range $\omega < \omega_p$ such that $\operatorname{Re}\varepsilon_1(\omega) < 0$. Wave Eq. (12) in such a case takes the form [11, 13].

$$\nabla^2\vec{E} - \frac{\varepsilon}{c^2}\frac{\partial^2\vec{E}}{\partial t^2} = 0 \tag{25}$$

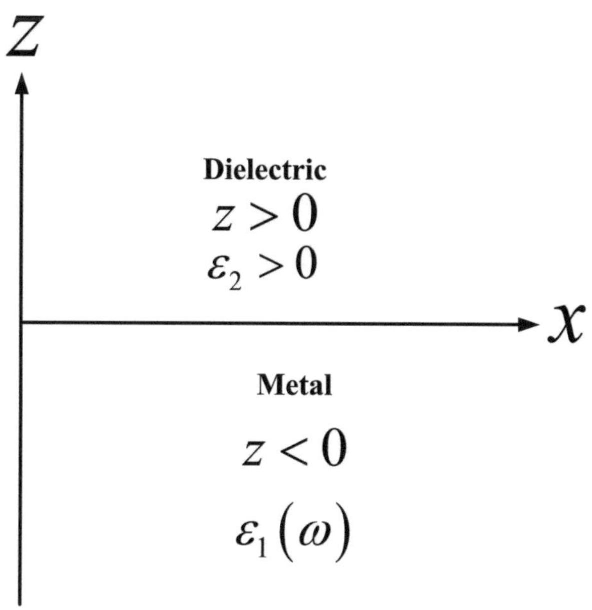

Figure 1.
Interface between a metal (z < 0) and a dielectric (z > 0).

where it is assumed that the permittivity ε in the both media is constant over distances of the order of magnitude of an optical wavelength and $\nabla \cdot \vec{D} = 0$ since the free external charges are absent [11].

The solution of the wave Eq. (25) in such a case is sought to be {11, 13].

$$\vec{E}(x,z,t) = \vec{E}(z) \exp i(\beta x - \omega t) \tag{26}$$

Here β is the propagation constant which is complex in the general cases. It can be shown that there exist two types of solution of the Maxwell's Eqs. (1)-(4) with different polarizations of the propagating waves [11, 13]. The first solution is the transverse magnetic (TM) or p modes with the field components E_x, E_z, H_y; the second solution is the transverse electric (TE) or s modes with the field components H_x, H_z, E_y [11, 13]. We start with the TM modes. The TM electric and magnetic fields in the regions $z < 0$ and $z > 0$ have the form, respectively [11, 13].

$$z < 0 \quad H_{1y} = A_1 \exp k_1 z \exp i(\beta x - \omega t) \tag{27}$$

$$E_{1x} = -iA_1 \frac{k_1}{\omega \varepsilon_0 \varepsilon_1(\omega)} \exp k_1 z \exp i(\beta x - \omega t) \tag{28}$$

$$E_{1z} = -A_1 \frac{\beta}{\omega \varepsilon_0 \varepsilon_1(\omega)} \exp k_1 z \exp i(\beta x - \omega t) \tag{29}$$

and

$$z > 0 \quad H_{2y} = A_2 \exp(-k_2 z) \exp i(\beta x - \omega t) \tag{30}$$

$$E_{2x} = iA_2 \frac{k_1}{\omega \varepsilon_0 \varepsilon_2} \exp(-k_2 z) \exp i(\beta x - \omega t) \tag{31}$$

$$E_{2z} = -A_2 \frac{\beta}{\omega \varepsilon_0 \varepsilon_2} \exp(-k_2 z) \exp i(\beta x - \omega t) \tag{32}$$

where $k_{1,2}$ are the components of the wave vector in the Z-direction perpendicular to the interface. The value of $1/|k_{1,2}|$ is the evanescent decay length of the fields (27)-(32) in the direction perpendicular to the interface. It defines the confinement of the TM waves. The boundary conditions of the continuity of the tangential field components H_y and E_x at the interface $z = 0$ have the form.

$$H_{1y}(z = 0^-) = H_{2y}(z = 0^+); \; E_{1x}(z = 0^-) = E_{2x}(z = 0^+) \tag{33}$$

Substituting expressions (27), (28), (30) and (31) into Eqs. (33) we obtain [11].

$$A_1 = A_2; \; \frac{k_2}{k_1} = -\frac{\varepsilon_2}{\varepsilon_1(\omega)} \tag{34}$$

On the other hand, substituting expressions (28), (29) and (31), (32) into wave Eq. (25) for the medium 1 and medium 2, respectively, we obtain the following dispersion relations [11, 13].

$$k_1^2 = \beta^2 - k_0^2 \varepsilon_1(\omega) \tag{35}$$

$$k_2^2 = \beta^2 - k_0^2 \varepsilon_2 \tag{36}$$

Here $k_0 = \omega/c$ is the absolute value of the wave vector of the propagating wave in vacuum. Substituting the second expression (34) into Eqs. (35), (36) we obtain the SPP dispersion relation [11, 13].

$$\beta = k_0 \sqrt{\frac{\varepsilon_1(\omega)\varepsilon_2}{\varepsilon_1(\omega) + \varepsilon_2}} \tag{37}$$

Analysis of TE modes shows that they cannot exist in the form of the surface modes [11, 13]. SPPs can exist only for the TM polarization [11, 13].

Analysis of dispersion relation (37) with the Drude dielectric function (22) shows that in the case of the mid-infrared or lower frequencies the SPP propagation constant β is close to k_0 [11]. As a result, SPPs can propagate over many wavelengths into the dielectric medium $z > 0$ [11]. These SPPs are called Sommerfeld – Zenneck waves.

In the case of large wave vectors, the SPP frequency tends to the surface plasmon frequency ω_{sp} given by [11].

$$\omega_{sp} = \frac{\omega_p}{\sqrt{1 + \varepsilon_2}} \tag{38}$$

The surface plasmon is an electrostatic mode which is a limiting case of SPP for $\beta \to \infty$ [11]. In the general case of the essentially complex dielectric function $\varepsilon_1(\omega)$ (22) the SPP energy attenuation length, or the propagation length $L = (2\mathrm{Im}[\beta])^{-1}$ which is usually between 10 and 100 μm for visible range optical frequencies and different types of the metal/ dielectric interfaces [11]. For instance, the SPPs at a silver/air interface for a vacuum wavelength $\lambda_0 = 450\,nm$ are characterized by the propagation length $L \approx 16\,\mu m$ and the evanescent decay length $1/|k_2| \approx 180\,nm$; the corresponding values for $\lambda_0 \approx 1.5\,\mu m$ are: $L \approx 1080\,\mu m$, $1/|k_2| \approx 2.6 nm$ [11]. Generally speaking, the lower SPP propagation length corresponds to the better SPP confinement [11]. The confinement of the optical field below the diffraction limit of $\lambda_0/2$ in the dielectric medium can be realized for optical frequencies $\omega \approx \omega_{sp}$ [11].

The SPP excitation at the metal/dielectric interface can be achieved by using the special phase-matching techniques such as a grating or prism coupling for the three-dimensional beams [11].

4. Nonlinear optical phenomena in Plasmonic nanostructures

In this section, we briefly discuss the peculiarities of the nonlinear optical effects in plasmonic structures. Plasmonics is a new field of photonics related to the interaction of light with matter in metallic nanostructures [12]. Plasmonics combines the capacity of photonics and the miniaturization of electronics because SPs and SPPs can confine light to subwavelength dimensions as it was mentioned above [11–13]. As a result, the effective nonlinear optical response can be enhanced significantly [3]. SPP crystals and waveguides, nano-antennas and plasmonic metamaterials can be used for the creation of optical responses by using the resonances of the individual units and their electromagnetic coupling [3]. The response of materials to an optical field is determined by its polarization $\vec{P}\left(\vec{r},t\right)$ which is given by [1–5].

$$P_i\left(\vec{r},t\right) = \varepsilon_0 \times \left[\chi_{ij}^{(1)} E_j\left(\vec{r},t\right) + \chi_{ijk}^{(2)} E_j\left(\vec{r},t\right) E_k\left(\vec{r},t\right) + \chi_{ijkl}^{(3)} E_j\left(\vec{r},t\right) E_k\left(\vec{r},t\right) E_l\left(\vec{r},t\right) + \ldots\right] \tag{39}$$

Here $i, j, k, l = 1, 2, 3$, $\chi_{ij}^{(1)}, \chi_{ijk}^{(2)}, \chi_{ijkl}^{(3)}$ are the linear, second-order and third-order optical susceptibilities, respectively [1–5]. In the general case, the optical susceptibility $\chi^{(k)}$ is the $(k+1)$ th–rank tensor [1–5]. In the centrosymmetric media $\chi^{(2)} = 0$, in homogeneous and isotropic media optical susceptibilities are scalar quantities [1–5].

The second and third-order nonlinear optical phenomena are the most important for applications [1–5]. The second-order response results in the wave-mixing effects such as the sum and difference frequency generation and second harmonic generation (SHG) where the incident frequency ω gives rise to the term with the frequency 2ω [1–5]. The third-order response contains the terms with the incident frequency ω, the new harmonics with the combination frequencies, and the third harmonic with the frequency 3ω [1–5]. The term with the incident frequency ω is caused by the Kerr effect where the refractive index n is modified by the optical field [1–5]. It is given by [1–5].

$$n = n_0 + n_2 I \tag{40}$$

where n_0, n_2 are linear and nonlinear refractive indices of the medium, respectively, and I is the optical field intensity. The optical Kerr effect results in all-optical switching and light modulation [3]. The local field $\vec{E}_{loc}\left(\omega, \vec{r}\right)$ enhancement at the metal/dielectric interface caused by the SPPs or LSPs excitation is characterized by the frequency-dependent local-field factor $L\left(\omega, \vec{r}\right)$ given by [3].

$$L\left(\omega, \vec{r}\right) = \left|E_{loc}\left(\omega, \vec{r}\right)/E_0(\omega)\right| \tag{41}$$

where ω is the plasmonic excitation frequency and $E_0(\omega)$is the incident field.

The surface–enhanced SHG, third-harmonic generation and FWM in nanoplasmonic structures have been demonstrated experimentally [3]. It has been shown that the SHG signal at the electrochemically roughened silver surfaces was increased by four orders of magnitude for a flat reference surface [3]. However, the SHG in such a case is diffused and incoherent being enhanced by LSP resonances of the nanoscale surface features [3].

The structured plasmonic surfaces for enhanced nonlinear effects represent metal gratings, arrays of different types of nanoparticles and nano-apertures, split-ring resonators [3]. For instance, FWM from a gold grating was 2000 times stronger than FWM from a flat film [3].

Controlling light with light in plasmonic nanostructures is based on LSP and SPP nonlinear effects. In such a case plasmonic structures are used for all-optical modulation or switching due to the enhanced Kerr-type nonlinearities [3]. For instance, the plasmon-enhanced nonlinear materials are metal nanoparticles and bulk materials doped with such nanoparticles [3]. Propagating SPPs can serve as the signal carrier [3]. The modulation and switching of SPPs in plasmonic waveguides can be achieved by changing of the real or imaginary part of the permittivity $\varepsilon_{1,2}$ by controlling light in the metal or dielectric at the interface shown in **Figure 1** [3]. Optically controlled dispersion in plasmonic waveguides can be realized by the Kerr nonlinearity in dielectric materials or by using the nonlinearity in a metal [3]. It has been shown that under certain conditions the bistability and modulational instability are possible in nonlinear plasmonic waveguides [3]. Plasmonic soliton-like excitation is also possible if the gain and nonlinearity are balanced correspondingly [3].

Consider now the plasmonic metamaterials [3, 7, 8]. They may be used for all-optical switching based on the plasmonic resonances of the split-ring resonators or nanorods and the electromagnetic coupling between these elements [3]. All-optical modulation in metamaterials can be realized by controlling the coupling strength between molecular excitons and plasmonic excitations [3].

Tunable and nonlinear metamaterials can be created by inserting of LCs into a metamaterial structure [7]. The optical fields and the bias electric field are simultaneously applied to a metamaterial with LC influence on LC reorientation [7]. Their interplay demonstrates a specific mechanism of electrically controlled optical nonlinearity in metamaterials [7]. For example, all-optical control of metamaterials with E7 NLC at the telecom wavelength of 1550 nm was investigated experimentally [7]. The integration of highly nonlinear LCs with plasmonic and metamaterials enables active switching and tuning of optical signals with a very low threshold [8]. Three general methods are combining the responses of highly nonlinear NLCs and plasmonics: (i) dissolving of nanoparticles in a bulk NLC cell; (ii) incorporating LC into nanostructures; (iii) chemical synthesis of nanoparticles with LC molecules [8]. Consider the case of a small volume fraction $f \ll 1$ of nano-spheres suspended in LC [8]. In such a case the effective optical dielectric constant ε_{com} of the composite material is given by [8].

$$\varepsilon_{com} = \frac{1+2f\gamma}{1-f\gamma}\varepsilon_{host}, \gamma = \frac{\varepsilon_{Au} - \varepsilon_{host}}{\varepsilon_{Au} + 2\varepsilon_{host}} \tag{42}$$

Here ε_{Au} is the permittivity of the Au nano-spheres described by the Drude model (22) and ε_{host} is the NLC permittivity. It has been demonstrated experimentally that a 100 μm thick sample of NLC L34 containing 0.5% Au nanoparticles provides the enhancement of the nonlinear absorption coefficients by about 250% or even more due to the inclusion of these plasmonic nano-particles into NLC cells [8].

5. Conclusions

We considered the interaction of the optical waves with free electrons in metals. The theoretical analysis in such a case is based on Maxwell's equations and the equation of motion of a free electron in an external electric field. As a result, plasmonic excitations emerge. The dielectric function of a free electron gas with plasmonic oscillations is described by the dielectric function (22) known as the Drude model. The longitudinal plasma oscillations excited at the plasma frequency $\omega = \omega_p$ are called the volume plasma oscillations, and the quasi-particles related to these oscillations are called the volume plasmons. SPPs may exist at the interface of the metal described by the Drude dielectric function and the dielectric medium. SPPs are the TM waves. They are characterized by the limited propagation length in the dielectric and the spatial confinement in the direction perpendicular to the propagation direction. The subwavelength confinement of the optical field is possible for SPPs. The nonlinear optical effects in plasmonic structures are significantly enhanced. The nonlinear optical phenomena such as modulation, switching, harmonic generation, FWM are possible in plasmonic nano-structures. Highly efficient all-optical switching and all-optical modulation can be realized in nonlinear plasmonic metamaterials based on nano-particles, split-ring resonators, and LCs.

Plasmonics plays an essential role in modern nonlinear optics. The nonlinear optical effects can be achieved at the reduced optical power due to enhanced effective nonlinearity in plasmonic nanostructures. The size of the nonlinear components can be scaled down which is important for the development of functional nanophotonic circuits. The ultrafast optical signal processing on the femtosecond scale can be realized due to the small response time of the plasmonic excitations.

Author details

Boris I. Lembrikov
Department of Electrical Engineering and Electronics, Holon Institute of Technology (HIT), Holon, Israel

*Address all correspondence to: borisle@hit.ac.il

References

[1] Shen YR. The Principles of Nonlinear Optics. Hoboken, New Jersey, USA: Wiley; 2003. p. 563. DOI: ISBN: 0-471-43080-3

[2] Moloney JV, Newell AC. Nonlinear Optics. Boulder, Colorado, USA: Westview Press; 2004. p. 436. DOI: ISBN: 0-8133-4118-3

[3] Kauranen M, Zayats AV. Nonlinear Plasmonics. Nature Photonics. 2012;**6**: 737-748

[4] Dombi P, Pápa Z, Vogelsang J, et al. Strong-field nano-optics. Reviews of Modern Physics. 2020;**92**:025003-1-025003-66

[5] Panoiu NC, Sha WEI, Lei DY, Li G-C. Nonlinear optics in plasmonic nanostructures. Journal of Optics. 2018; **20**:1-36

[6] Krasnok A, Alu A. Active nanophotonics. Proceedings Of IEEE. 2020;**108**:628-654

[7] Lapine M, Shadrivov IV, Kivshar YS. Colloquium: Nonlinear metamaterials. Reviews of Modern Physics. 2014;**86**: 1093-1123

[8] Khoo IC. Nonlinear optics, active plasmonics and metamaterials with liquid crystals. Progress in Quantum Electronics. 2014;**38**:77-117

[9] Khoo IC. Liquid Crystals. Hoboken, New Jersey, USA: Wiley; 2007. p. 368. DOI: ISBN: 978-0-471-75153-3

[10] Khoo IC. Nonlinear optics of liquid crystalline materials. Physics Reports. 2009;**471**:221-267

[11] Maier SA. Plasmonics: Fundamentals and Applications. Bath, United Kingdom: Springer; 2007. p. 223. DOI: ISBN: 0-387-33150-6

[12] Gavrilenko VI. Optics of Nanomaterials. Singapore: Jenny Stanford Publishing; 2020. p. 480. DOI: ISBN 978-981-4774-59-8

[13] Sarid D, Challener W. Modern Introduction to Surface Plasmons. New York, USA: Cambridge University Press; 2010. p. 371. DOI: ISBN 978-0-521-76717-0

Section 2

Nonlinear Nanophotonics

Chapter 2

Nanophotonics: Fundamentals, Challenges, Future Prospects and Applied Applications

Muhammad Aamir Iqbal, Naila Ashraf, Wajeehah Shahid, Muhammad Awais, Abdullah Khan Durrani, Khurram Shahzad and Mujtaba Ikram

Abstract

Nanophotonics encompasses a wide range of nontrivial physical effects including light-matter interactions that are well beyond diffraction limits, and have opened up new avenues for a variety of applications in light harvesting, sensing, luminescence, optical switching, and media transmitting technologies. Recently, growing expertise of fusing nanotechnology and photonics has become fundamental, arising outskirts, challenging basic experimentation and opportunities for new technologies in our daily lives, and played a central role in many optical systems. It entails the theoretical study of photon's interactions with matter at incredibly small scales, known as nanostructures, in order to prepare nanometer scale devices and accessories for processing, development, slowing down, influencing, and/or regulating photons through comprehending their behavior while interacting with or otherwise traveling via matter. This multidisciplinary field has also made an impact on industry, allowing researchers to explore new horizons in design, applied science, physical science, chemistry, materials science, and biomedical technologies. The foundations, nano-confinements, quantum manifestations, nanoscale interactions, numerical methods, and peculiarities of nonlinear optical phenomena in nano-photonics as well as projected nano-photonics consumption's in our cutting-edge world, will be covered in this chapter.

Keywords: nanophotonics, foundations, nano-confinements, quantum confinements, optical interactions, numerical simulations, nonlinear optical phenomena

1. Introduction

Nanophotonics is concerned with light-matter interactions at the nanoscale, which poses challenges to fundamental science while also opening the door to technological innovations. It encompasses the investigation of novel optical interactions, materials, manufacturing techniques, and models, as well as the exploration of organic and inorganic, or chemically manufactured structures such as holey fibers, photonic crystals, sub-wavelength structures, quantum dots, and plasmonics [1, 2]. Photonic medicine has become a rapidly emerging and theoretically incredible methodology for detection, disease prevention, and treatment. Because of the extremely fast rate of light modulation and the remote nature of optical procedures, light could be able to recover diagnostics, medications, and, unexpectedly, treatment course in a single theranostic procedure mixture of therapeutics and diagnostics, which provide clinical screening and therapy tracking [3]. It is concerned with the use of photonics in nanostructure media, when light is compressed down through nanometer scale volume and field enhancement effects emerge, resulting in new optical wonders that can be used to counter current advanced cutoff points and produce dominant superior photonic devices, which include a wide range of topics, such as metamaterials, quantum dots, quantum nanophotonics, high resolution imaging, plasmonics, and functional photonic materials. It has recently become a broadly recognized research field, and it will play an incredible role in the advancement of groundbreaking emerging innovations, ranging from high-efficiency solar cells to customized health tracking instruments that can detect the chemical structure of molecules at ultralow concentrations. Nanomaterials establish a substantial space in nanophotonics, and as we can see in this section and others to come, nanoscale optical materials span a wide variety of optical applications and have an incredibly diverse spectrum of nanostructure architecture. The optical properties of these nanostructures can be closely monitored by modifying them, allowing for the enhancement of one photonic function when presenting another photonic manifestation, and also the convergence of several functions to achieve multifunctionality [4].

2. Foundations of nano-photonics

2.1 Confinements approaches

Nanophotonics combines many important innovation thrust fields, including lasers, photovoltaics, biotechnology, photonics, and nanotechnology. Recently, growing expertise of fusing nanotechnology and photonics has become fundamental, arising outskirts, challenging basic experimentation. It can be divided into three types of confinement techniques: The first is to create nanoscale connections between light and matter by confining light to nanometer-sized dimensions far less than the light's wavelength. The approach that follows is to confine matter to nanometer range, restricting light-matter interactions to nanoscopic scales and characterizing the world of nanomaterials. The final method requires the nanoscale confinement of a photo-process through photochemistry or a light-induced phase transition, and it is used to fabricate photonic structures and functional units at the nanoscale. One method for confining light to a nanometer scale is to use near-field optical transmission, a model in which light is compressed via a metal-coated, tapered optical fiber and then exudes through a tip with an aperture far smaller than the wavelength of incident light. Different methods of confining the dimensions and producing nanostructures for photonic applications are used in nanoscale matter confinement; for example, nanoparticles with exceptional electronic and photonic properties. It's promising to hear that these nanoparticles are now being used in nanophotonic applications including UV absorbers that are used in sunscreen lotions. These nanoparticles may be composed of organic or inorganic materials, such as nanomers, having size-dependent optical properties as they are nanometer-sized oligomers (with a finite number of identical units) complexed with monomeric organic analogues, and polymers, which are long chain structures with a large number of repeating units. The field of "plasmonics" is made up of metallic nanoparticles that have an interesting optical reaction and an improved electromagnetic field. There are nanoparticles that can up-convert two absorbed IR photons into a visible UV photon, as well as quantum cutters that can down-convert an absorbed vacuum UV photon to two-visible UV photons. A photonic crystal is a hot field of nanomaterials that refers to a periodic dielectric structure with a repeated unit of the order of wavelength of light. Nano-composites are made up of phase-isolated nano-domains of at least two dissimilar materials on a nanometer scale. Each nano-domain in the nano-composite will give the bulk media a unique optical property. Controlling the flow of optical energy between various domains through an energy move (optical communications) is also possible. Nanolithography can be used to build nanostructures, which can then be used to fabricate nanoscale sensors and actuators using nanoscale photo-processes. The ability to confine photo-processes to all around characterized nano-regions, allowing structures to be fabricated in exact geometry and arrangement, is a key feature of nanofabrication. This section will illustrate the fundamentals of nanophotonics by describing the similarities and variations between photons and electrons, as well as confinement effects on photons and electrons caused by optical and electronic interactions at nanoscale range.

2.2 Photons and electrons: a comparison of their similarities and dissimilarities

Photons and electrons are subatomic elementary particles that can function as both particles and waves. Electrons are negatively charged subatomic particles with the smallest mass, while photons are massless quanta of energy that leads to electromagnetic radiations. The intrinsic angular momentum of an electron is a half integer of ħ (spin = 1/2), indicating that it is a fermion. As a result, if more than one electron occupies the same space, each electron's properties should be unique and conform to Fermi-Dirac statistics. The Pauli's exclusion principle depicts the strong interaction of electrons (fermions). Photon, on the other hand, is an elementary particle with both electric and magnetic fields governed by Maxwell's equations, as well as an inherent angular momentum of integer magnitude of ħ (spin = 1), indicating that it is a boson. According to Bose-Einstein statistics, photons do not associate with other photons, so more than one photon can occupy a single quantum state. There are two ways in which photons differ from electrons: (I) Photons are vector fields (light has the ability to be polarized), while electron wavefunctions are scalar; and (II) Electrons have charge and spin, while photons do not.

Atoms are made up of nuclei (neutrons and protons) surrounded by electrons. Since light (photons) that interacts with nuclei need a lot of energy (gamma rays), hence X-ray to infrared light only interacts with electrons, regulating photon/electron interactions. Light energizes an electron cloud with one particle, which emits a photon, which interacts with another electron cloud in transparent materials, while light can be absorbed and emitted in opaque materials by a resonant electron and photon connection. Quantum representation shows that electrons and photons are analogous and have several characteristics. **Table 1** summarizes the similarities in features of electrons and photons.

2.3 Confinement of photons and electrons

Photon and electron propagation can be dimensionally restricted by reflecting or backscattering these particles in spaces of varying interaction potential along their propagation path, thereby confining their propagation to a single direction or range of directions. According to classical mechanics, electrons and photons are fully confined in confinement areas. Since the energy of electrons trapped within potential energy limits is less than the potential energy due to the boundary, they remain

Characteristic	Photons	Electrons
Wavelength	$\lambda = \frac{h}{P} = \frac{c}{\upsilon}$	$\lambda = \frac{h}{P} = \frac{h}{mv}$
Wave Equation	$\{\nabla \times \frac{1}{\varepsilon(r)} \nabla \times\}\, B(r) = (\frac{\omega}{c})^2 B(r)$	$\hat{H}\Psi(r) = -\frac{\hbar^2}{2m}\nabla.\nabla + V(r)\Psi(r) = E\,\Psi$
Free-Space Propagation	Plane wave $E = (\frac{1}{2})E^0\,(e^{ik.r-\omega t} + e^{ik.r+\omega t})$ K is a wave vector and a real quantity	Plane wave $\Psi = c\,(e^{ik.r-\omega t} + e^{ik.r+\omega t})$ K is a wave vector and a real quantity
Interaction Potential	Dielectric constant (refractive index)	Coulomb interactions
Propagation in Classically Forbidden Zone	Photon tunneling (evanescent wave) with imaginary wave vector k and exponentially decaying amplitude	Electron tunneling with exponentially decaying amplitude (probability).
Localization	Strong scattering caused by massive dielectric constant variations (Photonic Crystals)	Big differences in coulomb interactions cause strong scattering (Electronic Semiconductor Crystals)
Cooperative Effects	Interactions between nonlinear optical systems	Many-body correlation Superconducting Cooper pairs Bi-exciton Formation

Table 1.
An overview of similarities in characteristics of electrons and photons.

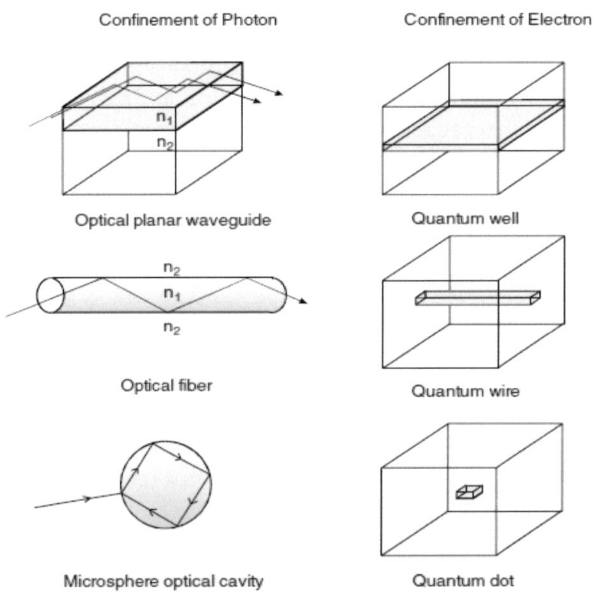

Figure 1.
Confinement of photons and electrons in different dimensions with propagation along z-axis [6].

completely enclosed within the walls. The wave picture of photons and electrons, on the other hand, does not indicate this. Photons can be used to envision confinement by collecting light in an environment with a high refractive index or high surface reflectivity, whereas a waveguide or cavity resonator can serve as a confining field [5]. The **Figure 1** depicts the classic picture of light direction (trapping) due to total internal reflection using a beam line. The confinement in a planar waveguide is only along the vertical x-direction, while the propagation direction is along the z-axis whereas the confinement of a fiber or a channel waveguide is in the x and y directions. A microsphere is a three-dimensional representation of an optical medium that confines light in all directions by comparing the refractive index of the leading and surrounding mediums. As a result, the disparity n1/n2 functions as a scattering potential, obstructing light propagation. Light functions as a plane wave with a continuous propagation constant, similar to the propagation vector k in free space; the electric field circulation has a distinct spatial profile in the direction of propagation (z axis) and in the confining direction [6].

2.4 Confinement of optical interactions at the nanoscale

Numerous geometries will confine the electric field associated with a photon (electromagnetic field) to initiate optical interactions at the nanoscale, including axial and lateral localization approaches that can be used to reduce the optical field on the nanoscale, as seen schematically in **Table 2**.

2.4.1 Axial nanoscopic localization

This technique employs the evanescent wave and surface plasmon methodologies, which are discussed further below.

I.Evanescent Wave

In the field of optics, evanescent waves, are oscillating electric and/or magnetic fields that may not propagate like electromagnetic waves but whose energy is spatially amassed near the source rather than moving waves of

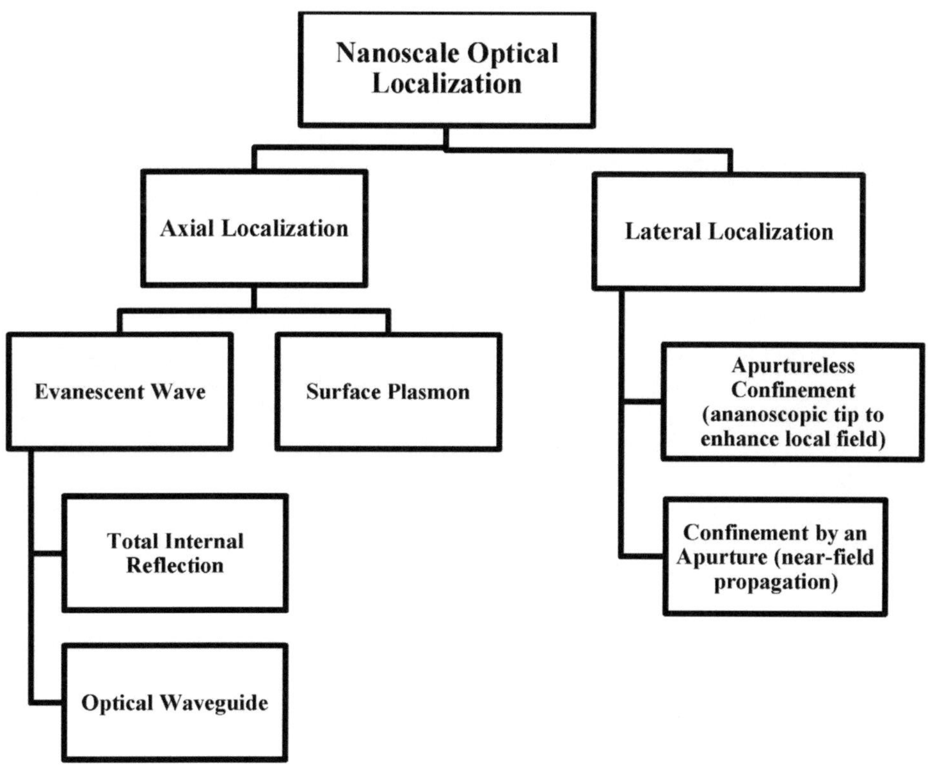

Table 2.
An overview of confining optical interactions at nanoscale.

gradually decaying field intensity in a specific spatial direction that are often encountered. They also do not contribute to energy transfer in that direction; despite the fact that the Poynting vector (averaged over one oscillation cycle) can have non-zero segments in various ways. Evanescent waves can also occur in various types of waves, such as sound waves and quantum-mechanical waves. There are likewise situations where a light field can be disintegrated into an evanescent and a propagating part. A light field may also be disintegrated into an evanescent and a propagating element under some circumstances. As a waveguide surface reaches a lower refractive index medium, an evanescent wave is formed, which decays exponentially in the axial direction (away from the waveguide) and this evanescent field has a magnitude of 50–100 nm that can be used to cause nanoscale optical interactions as well as evanescent wave excitation for high near-field fluorescence sensing [7]. The coupling of two waveguides by an evanescent wave is another example of nanoscale optical interaction, as seen schematically in **Figure 2(a)**, in which photons are launched from one waveguide to another, and these evanescent wave-coupled waveguides can be used as directional couplers in an optical communication network to relay signals. It has also been proposed that evanescent wave-coupled waveguides be used for sensor applications, where sensing induces a transition in photon tunneling from one waveguide channel to the next. There are optical sensors for specific material species where one can experiment with how a light field, which is essentially directed, such as in a glass structure, is connected to an evanescent field and atoms or molecules in that area will be able to interact with the light field; for example, after being excited by light, they could emit fluorescence [6].

Total internal reflection, which involves the scattering of light through a prism with a refractive index of n1 to an atmosphere with a lower refractive index of n2, is just another illustration of a geometry that produces an evanescent wave [7, 8]. As the incidence angle is small enough, light refracts and passes through the second medium to some degree, such that as the incidence angle reaches a critical angle, the light ray is reflected from the interface, as seen in **Figure 2(b)**.

II.Surface Plasmon Resonance (SPR)

At its most basic form, the Surface Plasmon Resonance (SPR) technique is an extension of the evanescent wave interaction depicted above, in which the waveguide or a metal–dielectric interface replaces the prism. Electromagnetic waves that propagate across the interface between a dielectric organic material and a metal film are known as surface plasmons [9, 10]. Since surface plasmons propagate as a wavevector with a specific frequency band in a metal film, no light may travel across any media, and no direct excitation of surface plasmons can be achieved immediately. The most popular approach for producing a surface plasmon wave is attenuated absolute reflection (ATR).

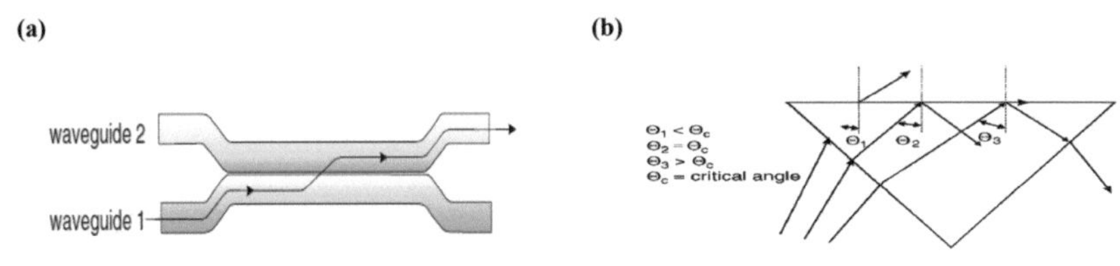

Figure 2.
(a) Evanescent wave-coupled waveguides (b) principle of total internal reflection [6].

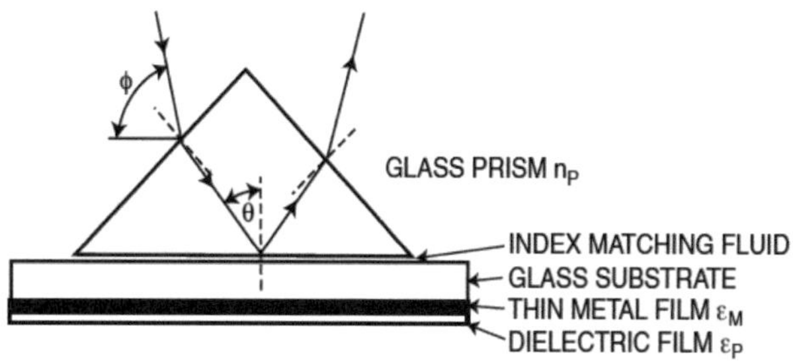

Figure 3.
Excitation of Plasmons using Kretschmann geometry [9].

Figure 3, depicts the Kretschmann structure of ATR, which is commonly used to excite surface plasmons [9].

2.4.2 Lateral nanoscopic localization

A near-field geometry in which the illuminated sample is located within a fraction of the wavelength of light emitted by the source or aperture, can be used to achieve lateral nanoscale light confinement [8, 10] and an electric field propagation around a nanoscopic system generates spatially localized optical interactions. Further, owing to the virtual values of wavevector-like characteristics, the distribution of the spatially localized electric field has a significant evanescent form, that is, it decays exponentially. To study near-field geometry, a near-field scanning optical microscope (NSOM) with aperture based and appertureless configurations may be used, and an aperture-based NSOM uses a submicron-sized 50–100 nm aperture, near to the tapered optical fiber's opening tip, utilized to keep light confined, while an apertureless structure maximizes the local field by a nanoscopic metal tip or a metallic nanoparticle [11].

2.5 Nanoscale confinement of electronic interactions

This section focuses on a few basic examples of nanoscale electronic interactions that cause drastic changes in the optical properties of a material. The results of confining electronic interactions at the nanoscale involve a variety of factors that control them, which are discussed below.

2.5.1 Quantum confinement effects

Quantum confinement alters the optical properties of semiconductors in a number of ways, and these variations have practical applications. As a semiconductor's length is limited to a similar order of exciton radius, i.e. a few nanometers, hence the quantum confinement effect happens. The dimensionality of electrons is reduced by confining them to a thin semiconductor surface, resulting in a dramatic increase in exciton properties and behavior. QuantumDots, Quantum Wires, and Quantum Wells are the three kinds of quantum confined structures that have been described so far based on the function of confinement [6]. The optical transformations caused by different confinements are mentioned in **Table 3**.

2.5.2 Quantum-confined stark effect

The quantum-confined stark effect explains how an applied electric field affects energy levels and thus optical spectra. Quantum-confined devices exhibit major

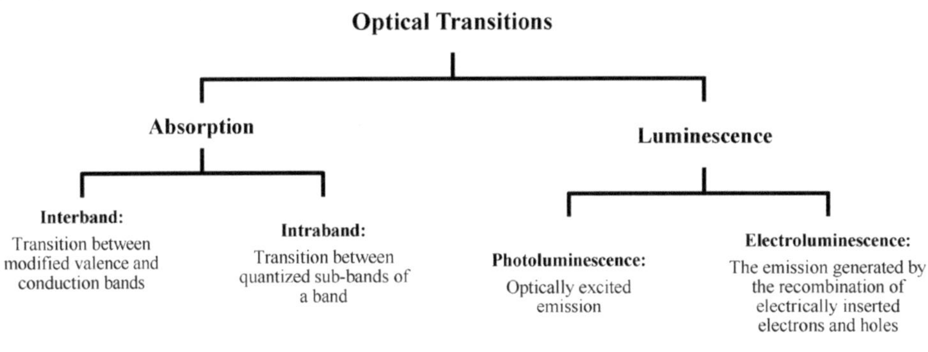

Table 3.
Optical transition under the effect of quantum confinement of electrons at nanoscale.

variations in their optical spectra where an electric field is applied in the direction of confinement [6]. The most important embodiments of quantum confinement effects are mentioned in the **Table 4**.

2.5.3 Dielectric confinement effect

The effects of dielectric confinement can be analyzed by adjusting the restricted semiconductor region's dielectric constant and the confining potential barrier that surrounds it. Since the potential barriers created by compositional changes in a quantum well do not create a large difference in the dielectric constant, this effect is always dispersed. Depending on the processing and fabrication process, a quantum rod, quantum dot, or quantum well may be inserted in another semiconductor or dielectric, such as glass or polymer, and these quantum designs may be characterized by organic ligands, dispersed in a liquid, or merely encompassed by air. When the surrounding medium's dielectric constant is ultimately smaller than that of the confined semiconductor system, such media will exhibit a greater shift in the dielectric constant and significant changes of optical properties due to dielectric confinement [12, 13].

2.6 Nanoscopic interaction dynamics

Controlling the dynamics of local interactions inside nanostructure media that enhance complex radiative transitions is a model accomplished using a nanocrystal host environment with low-frequency phonons to reduce multi-phonon relaxation of excitation energy in order to increase emission efficiency of rare-earth ion's. Since rare-earth ion's electronic transitions are highly susceptible to nanoscale interactions, manipulating the existence of electronic interactions requires only a nanocrystalline media, allowing for the use of a glass or plastic medium in a wide range of device technologies. As discussed further below, nanoscale electrical contacts between two electronic centers result in novel optical transitions and improved optical communications [6]. This sub-section will provide a glimpse of interaction dynamics occurred at nanoscale range.

2.6.1 New cooperative transitions

Two adjacent species may interact through a series of ions, atoms, or molecules to generate optical absorption bands or to allow novel multiphoton absorption

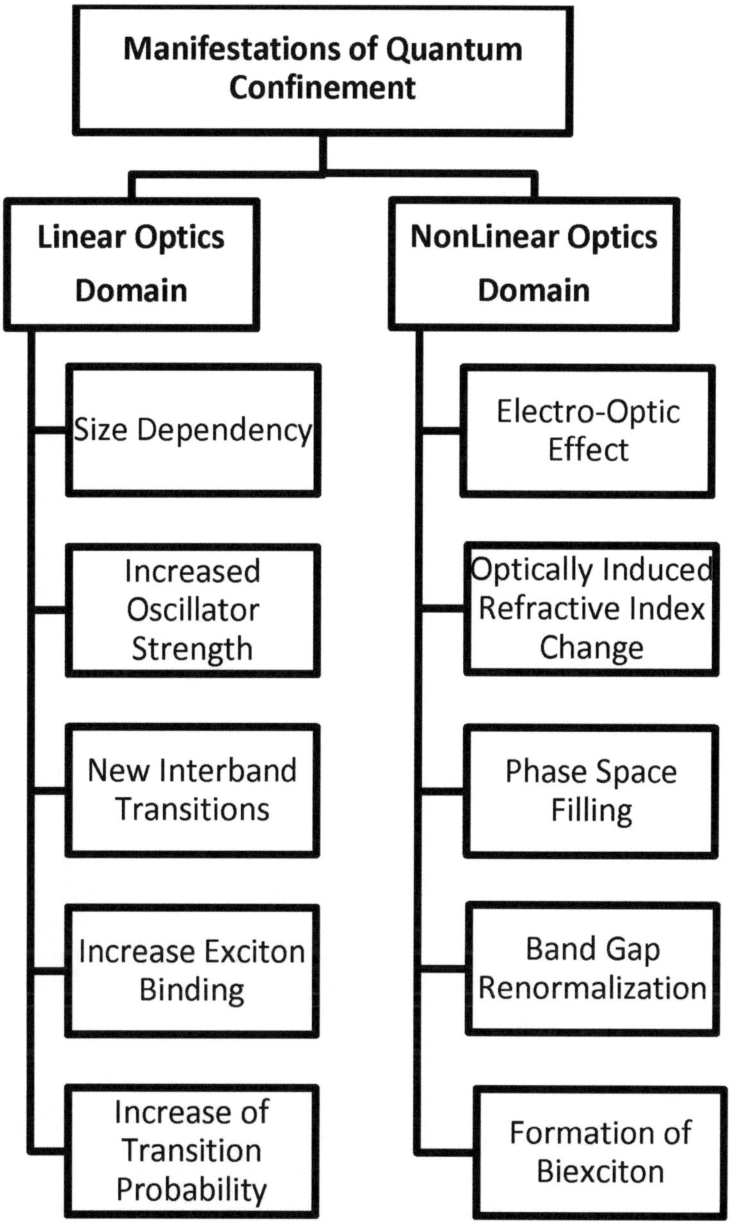

Table 4.
Manifestations observed under the effect of quantum confinement.

processes. The production of biexcitons in a quantum-confined system or a semiconductor yields new optical absorption and emission of lower energy than two individual excitons, and the energy difference correlates to the excitons binding energy. The joining of many excitons to form a multiexciton or exciton string has also been suggested as an expansion of the biexciton principle. In the context of a molecular structure, an example will be the creation of multiple aggregates, known as a J-aggregate of dyes, which is a head-to-head alignment of various dye dipoles [14]. The other kind of nanoscale electronic interaction takes place whenever an electron-donating group or molecule comes into close proximity to an electron-withdrawing group or molecule in nanoscopic space, new optical transitions occur. An organometallic structure is an example of association of an inorganic ion to a wide number of organic groups. Novel optical transformations engaging charge transfer from metal to ligand (MLCT) or, in some situations, light absorption causes a reverse charge transfers, provided by these types of organometallic structures [7]. An intermolecular organic donor (D)–acceptor (A) complex, is another example, which produces charge transfer species D + A in the excited state despite the fact that the constituents D and A are colorless and have no visible absorption. These charge-transfer complexes have clear visible color resulting from recent charge transfer transitions in the visible spectrum [6].

2.6.2 Cooperative emission

Cooperative emission is a manifestation of electronic interactions, which happens when two nearby centers are electronically excited within nanoscopic distances, resulting in a higher-energy photon being released as a result of a simulated state of the pair centers. This mechanism is seen in rare-earth ions, resulting in an up-converted emission of a higher energy photon than the excitation energy of actual ions [6]. The interaction reappears as two neighboring ions are separated by just a few nanometers and based on the form of electronic excitation of the individual ions, the interaction between these two ions can be multipole–multipole or electron exchange. It's worth noting that the emission comes from a virtual level rather than the ion pair's physical level.

2.6.3 Nanoscale electronic energy transfer

Additional electrical energy emitted by an optical transition occurred due to nanoscopic interaction dynamics may be transmitted from one center (atom, ion, or molecule) to the next, typically on a nanoscopic level, though long-range transfer of energy is also possible. This transfer of electrical energy does not require the flow of electrons, but rather the transfer of excess energy. As a result, one center of this mechanism has excess energy in an excited electronic state and serves as an energy donor by passing the excitation to an acceptor, resulting in an electron in the energy donor group becoming excited and returning to the ground state, whereas an electron in the energy acceptor group becomes excited. Exciton migration occurs as a result of interactions between energetically related centers, either coherently by a series of closely spaced levels creating an exciton band or by hopping an electron–hole pair from one core to another incoherently [6]. Fluorescence resonance energy transfer (FRET) is also another form of energy transfer that happens when two different kinds of molecules interact. Fluorescence from the energy acceptor can be detected by optically exciting a molecule to a higher electronic state, and this type of transition is frequently observed with two fluorescent centers separated by a few nanometers. FRET is a widely used bioimaging technique for probing nanoscale interactions between cellular components, such as protein–protein interactions [7]. Throughout this context, one protein may be identified with a fluorescent dye which, once electronically excited by light, behaves as an energy donor. If two proteins are well within 1–10 nm of one another, the other protein is identified as an energy acceptor, and will absorb energy when the two proteins are within this nanoscopic range of distances. This type of energy transfer occurs often in a dipole–dipole interaction with a distance dependency between the energy donor and acceptor. To maximize FRET activity, the donor's emission spectrum and acceptor's absorption spectrum may have significant spectral overlap [6].

3. Computational nano-photonics: standard problems and numerical methods

Nanophotonic structures are diverse, and their physical characteristics are related to one or more observables, that are required for their comprehension and design, including the distribution of an electric field inside a photonic cavity or a nanoparticle's scattering cross section. Depending on the extent of discretization, computational methodologies have been proposed to simulate the functional behavior of complex nanophotonics structures that do not have an analytical solution. This can be translated as approximating a physical problem with a set of appropriate analytical functions defined on a finite or infinite domain. Many common approaches that have been thoroughly studied in the context of computational electromagnetics [15, 16], with some of them have been quantitatively compared or validated to analytical or experimental findings in the field of nanophotonics [17–22]. However, finding functional cases that can equally compare the various numerical approaches remains a challenge. The methodology used to derive the approximated problem has a significant impact on the numerical method's strengths and disadvantages, including the kinds of challenges it is best suited for. Additionally, in comparison to calculating time and memory constraints, the performance of a numerical system requires a variety of other factors, including ease of execution, discretization complexity, and flexibility. Light propagation, localization, scattering, and multiscale problems are known as four types of problems that form the foundation for modeling nanophotonic systems and they can be handled with two categories of versatile approaches: integral (SIE, VIE) and differential (FE, FDTD, and hybrid FD/FE) methods. Time domain methods, like DGTD and FDTD, have now been described as being particularly suitable to light propagation problems, while the FE and SIE approaches are especially effective for problems involving light localization and finally, SIE and VIE based techniques, in addition of RCWA (unique to specific geometries), are well-suited to light scattering problems [23].

3.1 Finite differences in time domain (FDTD)

Due to the ability to tackle a wide range of problems, FDTD approach is one of the most common methods in nanophotonics [24]. A staircase approximation is used in this technique to define the volumes of the nanostructure, superstrate, and substrate that have both space and time discretized, and finite difference quotients are used to replace both spatial and temporal derivatives of Maxwell's curl Equations [15, 25]. The famous algorithm proposed by Yee [26] is an example of FDTD method in addition to numerous other approaches [15, 24, 25].

3.2 Finite elements (FE)

The finite elements (FE) method is another widely used differential technique in nanophotonics for calculating an effective frequency-domain electromagnetic field. The finite elements in time-domain (FETD) approach and the Discontinuous Galerkin time-domain (DGTD) method are two hybrid strategies that rely on the inside of each element, the DGTD approach explicitly solves Maxwell's equations and connects them to a quantitative flux [27, 28]. In a similar way to FDTD, the central difference principle can be used to discretize in the time domain. This approach allows for the use of higher level expansion and checking functions, as well as the local solution of equations in every element, resulting in high accuracy by combining high-order FE precision with a time-domain classification, allowing it to be used for large structures [29].

3.3 Volume integral methods

Volume integral (VIE) methods are used to determine the electric field on a smaller scale by modifying Maxwell's equations to integral form, and the numerical equations associated with these integral solutions can be obtained in a variety of different ways. The discrete dipole approximation (DDA) is a common implementation of the vectorial wave equation that can be used to derive numerical equations [30, 31].

3.4 Surface integral methods

Surface integral methods restrict an electromagnetic scattering problem with open boundary conditions to the surface limits of the substance. As a consequence, commonly used methods to piecewise homogeneous media, such as the boundary element method (BEM) [32] and the SIE method [20, 33], are best suited. The SIE process, like the FE and VIE methods, will specifically treat dispersive materials as a frequency-domain method [34].

3.5 Other methods

The hybrid differential–integral approaches are particularly useful in dealing with inhomogeneous scatters or anisotropic [27], and are an alternative to the integral and differential methods to solve Maxwell's equations. Brief overview of other methods used for simulation of nanophotonics includes;

3.5.1 Finite integration technique

The finite integration technique (FIT) is an integral form discretization scheme of Maxwell's Equations [35]. Unlike integral and differential approaches, which may cover a wide variety of structures, few other methodologies are considered to be very effective for unique conditions and geometries since they depend on electromagnetic field expansions on basis functions with specific symmetries [23].

3.5.2 T-matrix system

The incident and scattered electromagnetic fields are extended on a series of spherical basis functions in the T-matrix system, with boundary conditions imposed at the interfaces of the different materials. This approach excels at measuring the scattering of spherical and/or quasi-spherical particles [36] and extended to plasmonics, layered particles as well as particle-substrate interactions with applications in sensing, plasmonic trapping, and SERS [37, 38].

3.5.3 Multiple multipole approaches

The generalized multipole method is a semi-analytical technique, also known as the multiple-multipole approach that is used to extend the electromagnetic field in multipoles, allowing it to handle a wide variety of symmetries [39, 40]. In this method, only the domain boundaries are discretized and no integral is numerically solved. Multiple elastic scattering of multipole expansions disintegrates dispersed fields to multipoles with respect to centers near each cluster object, and this occurs until cluster convergence [41, 42].

3.5.4 RCWA approach

The RCWA uses intermittent grating systems to diffract electromagnetic waves. The periodic permittivity and electromagnetic field are expanded as Fourier series, and material boundaries are improved with boundary conditions [43], resulting in poorly produced abrupt surfaces with a large number of Fourier harmonics. This method is highly reliable for computing far-field reflection, propagation coefficients, and diffraction orders, while calculating the near field at material boundaries remains challenging [44].

4. Peculiarities of nonlinear optical phenomena in nano-photonics

Nonlinear Optics is the analysis of the mechanisms that occur as light manipulates a material's optical properties. Franken et al. [45] discovered second-harmonic generation soon after revealing the first working laser by Maiman in 1960 [46], and it is widely regarded as the start of the field of nonlinear optics. The discovery of saturation results in the luminescence of dye molecules identified by G. N. Lewis et al. [47] is the earliest known example to the author's. This section will explore the peculiarities of nonlinear optical phenomena in nanophotonics by including an overview of active nanophotonic devices, gain materials, plasmonic nanostructures, metamaterials, quantum dot lasers, and optical amplifiers.

4.1 Active nano-photonic devices

Active materials have recently appeared as a potential loss compensation technique, first in nanoparticles, metamaterials, and plasmonic waveguides, and then in novel functionalities such as signal amplification and lasing in the field of nanophotonics [48]. It's important to note that in nanophotonics, the word "active" refers to the manipulation of material properties like refractive index in phase-change materials to regulate or reorganize plasmon propagation. Nanolasers and surface plasmon amplifiers have piqued the attention of researchers since they enable the theory of coherent stimulated emission to be applied to the diffraction limit and beyond. The notion of a SPASER [49], acronym of "surface plasmon amplification by stimulated emission of radiation", was originally designed to amplify oscillating localized surface plasmons (LSP's), inside metal nanoparticles and it was eventually expanded to add traveling surface plasmon-polaritons (SPP's). Aside from their significance in optoelectronic and all-optical data processing, they are also used in sensing, biological and super resolution imaging [50].

4.2 New gain materials

Gain materials and parametric amplification through nonlinear effects are the key approaches for achieving optical gain in active systems. The standard active nanophotonic device scenario is depicted in **Figure 4**, in which a lossy resonator is surrounded by an active material and a bulk gain material may be used to explore this arrangement such as halide perovskite, chromophore, QD's scattered with metallic and/or dielectric nanoparticles or a gain material layer surrounding a nuclei core [50]. Material nonlinearity has been successfully used in metamaterials [51]

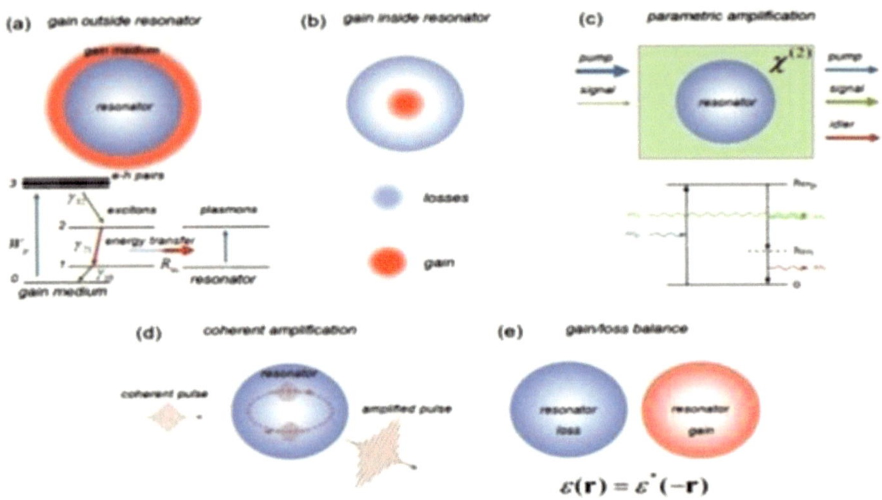

Figure 4.
Multiple approaches used for representation of active nanophotonics [50].

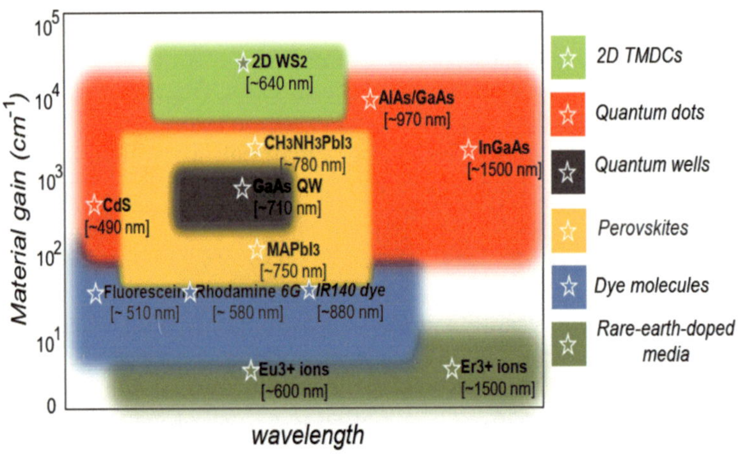

Figure 5.
Material gain parameter achieved for rare earth doped and other materials [50].

and nanostructures [52] for loss correction using optical parametric amplification (OPA), while coherent amplification, which is based on pulse amplification within a cavity through positive interference, may provide gain and has recently been suggested for loss compensation in plasmonics [50]. The achievable material gain parameter of prospective materials for nanolasers including transition metal dichalcogenides (TMDCs), quantum dots, quantum wells and perovskites is shown in **Figure 5** while **Figure 6** depicts an overview of new materials for nanolasers.

4.3 Nonlinear optics in plasmonic nanostructures

Metallic plasmonic structure, such as nanoparticle, optical diffraction grating, and nano-aperture, is one well-known and commonly studied method to vastly enhance the efficiency of optical near-field as well as nonlinear optical interactions at the nanoscale, whereas their role is threefold in nonlinear optics; (i) to increase the effect of nonlinearity (ii) to reduce the size of nonlinear components (iii) to have an ultrafast response time, allowing optical signals to be manipulated on

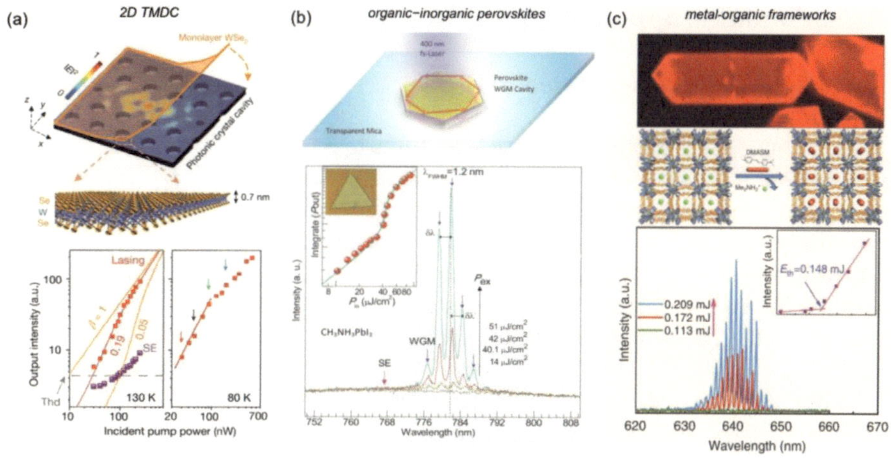

Figure 6.
An overview of new materials for nanolasers [50].

femtosecond timescales [53]. In plasmonics, metal dielectric interfaces support surface plasmon polaritons (SPP's) [54], which are p-polarized strongly scattered surface waves associated with oscillations of free electrons in metals, powerful spatial confinement and enhancement in the electric field occur at the interface between the two media following the production of these localized or propagating modes. Owing to the near confinement of the optical field of SPP's, surface effects at metal dielectric interfaces are especially sensitive to variations in the shape of plasmonic nanostructures and the dielectric properties of the embedding optical media. Surface plasmon-polariton modes have been suggested to limit the scale of lasers to sub-wavelength dimensions with the increase of plasmonics. Population inversion of emitters (QD's, fluorophore's) and feedback generated by plasmonic resonant structures are used to achieve lasing in such plasmonic-based structures. The development of new types of nano-resonators and active materials led to the development of a family of nanolasers that drew a lot of interest from the nanophotonics world [48, 50].

4.4 Nonlinear plasmonic metamaterials

Plasmonic metamaterials have potential prospects for all optical switching by tailoring the constituent's plasmonic resonances and the electromagnetic coupling between them. Both individual plasmonic resonances and their interactions can be altered as a function of variation in refractive index of the embedding dielectric or substrate, resulting in better nonlinear reactions. By controlling the binding power of molecular excitons, plasmonic excitations will result in effective all-optical modulation, and plasmonic metamaterials give a novel way to increase nonlinearity using epsilon-near-zero regime 'nonlocal' effects [53].

4.5 Quantum dot lasers and optical amplifiers

Quantum Dot (QD) based semiconductor optical amplifiers (SOA's) demonstrate ultrafast gain dynamics and pattern effect free amplification, both theoretically and experimentally [55]. Mode-locked (ML) lasers can be used as optical comb generators for high-frequency applications including time domain multiplexing because of its low alpha factor and wide spontaneous emission range of the quantum dot gain medium [56].

5. Nanophotonics: applications

5.1 Biomaterials and nanophotonics

Material engineers found biological structures to be a fertile ground, allowing for the creation of innovative nanotechnologies for a broad range of applications. Bioprocesses generate nearly flawless nanostructures that are biodegradable and can be used for efficient solar energy harvesting, filtering, high-density data storage, low-threshold lasing, and optical switching. For a wide range of photonic features, both active and passive nanophotonics systems can be employed in a wide range of biomaterials [6]. The four types of biomaterials are as follows;

i. Bio-derived materials have unrivaled molecular-level power for fine-tuning properties at both the single-molecule and bulk scales. Interdisciplinary approaches are needed for the design and engineering of bio-derived materials that use proteins, sugars and lipids as building blocks.

ii. Bio-inspired materials are synthetic nanostructures, produced by mimicking natural biological material synthesis processes using biological principles, whereas, biomimicry is a growing discipline that focuses on creating multifunctional hierarchical materials and morphologies that resemble nature and can be used as a light harvesting dendritic framework.

iii. Bio-templates are natural nanostructures with suitable morphologies and surface interactions that can be used as models to create multiscale and multicomponent photonics materials that provide reinforcing sites for photonic active structures to self-assemble.

iv. Bacteria bioreactors for metabolically engineering photonic polymers, where metabolically engineered materials are those generated by manipulating the naturally occurring bacterial biosynthetic mechanism to manufacture a family of helical polymers with a broad variety of optical properties.

5.2 Nanophotonics for biotechnology and nanomedicine

Nanophotonics has a wide range of uses in biomedical science and technology, including nanomedicine applications for light-guided and light-activated treatment, and also studies of the fundamentals of interactions and processes at the single cell/molecule level. Nanomedicine is a developing discipline that focuses on the use of nanoparticles in the creation of novel noninvasive diagnostics for early disease detection, as well as for promoting selective drug distribution, treatment efficacy, and real-time drug tracking. A thorough understanding of drug–cell interactions, with an emphasis on molecular modifications at the single-cell level caused by the pre-onset state of a disease, can be used to develop "personalized" clinical treatments for disease control focused on molecular identification. Using optical methods, nanophotonics helps one to map drug consumption, clarify the cellular mechanism, and track subsequent cytosolic interactions. For this reason, biosensing, bioimaging, and single-cell biofunction trials utilizing optical probes are particularly beneficial. Light-guided and light-activated treatments have made significant progress in the field of nanomedicine-based molecular disease identification. Nanoparticles are now

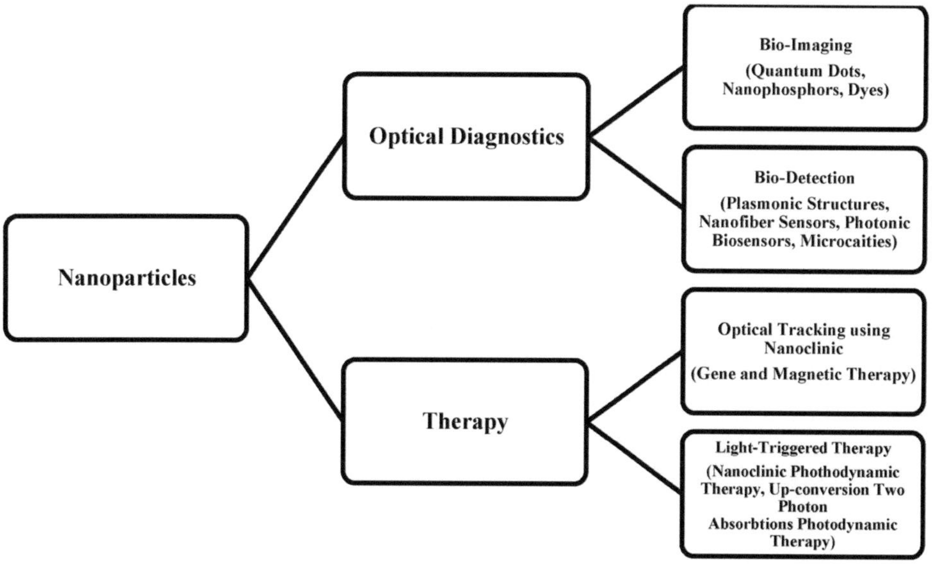

Table 5.
A glimpse of optical nanomaterials potential applications.

equipped with optical probes, advanced carrier groups, and light-activated therapeutic agents capable of guiding the nanoparticles to diseased cells or tissues to allow selective drug delivery and real-time drug efficacy monitoring [6].

5.3 Optical nanomaterials

At the moment, the most well-established applications for optical nanomaterials are in very low-tech applications such as sunscreen lotions and optical coatings, although many high-tech industrial solutions have recently emerged, including signal processing and photonic crystals for complex optical circuitry, as well as sensors for detecting and responding to chemical and biological challenges [6]. Several new companies have emerged to produce next-generation solar cells based on the nanoparticle principle, but they are still in the early stages of growth. Nanoparticles have also been shown to be useful in optical diagnostics, light-activated treatments, as well as in optical communication initiatives [6]. **Table 5** illustrates graphical applications of nanostructures in optics.

6. Conclusions

The nanoscale interaction dynamics and confinement of photons and electrons were depicted in this chapter, which served as fundamentals of nanophotonics. The foundation for modeling nanophotonic structures, as well as the methodologies used to tackle issues at the nanoscale regime, such as light propagation, scattering, localization, and multiscale problems, have been frequently discussed in terms of two forms of versatility. Nanophotonic structures are used in a wide range of applications, including nanomedicine, biomaterials, biotechnology as well as optical diagnostics, and they are the spark for the next forefront of economic development.

Author details

Muhammad Aamir Iqbal[1*], Naila Ashraf[2], Wajeehah Shahid[3], Muhammad Awais[4], Abdullah Khan Durrani[5], Khurram Shahzad[5] and Mujtaba Ikram[5]

1 School of Materials Science and Engineering, Zhejiang University, Hangzhou, People's Republic of China

2 Department of Mathematics, Government College University (GCU), Lahore, Pakistan

3 Department of Physics, The University of Lahore (UOL), Lahore, Pakistan

4 Zhejiang Provincial Key Laboratory of Electrical Machine Systems, College of Electrical Engineering, Zhejiang University, People's Republic of China

5 Institute of Chemical Engineering and Technology (ICET), University of the Punjab, Lahore, Pakistan

*Address all correspondence to: maamir@zju.edu.cn

References

[1] Ohtsu M, Kobayashi K, Kawazoe T, Yatsui T, Naruse M. Principles of nanophotonics. 2008;**6**.

[2] Shen Y, Prasad PN. Nanophotonics: a new multidisciplinary frontier. Applied Physics B. 2002;**74**:641-645.

[3] Conde J, Rosa J, Lima JC, Baptista PV. Nanophotonics for molecular diagnostics and therapy applications. International Journal of Photoenergy. 2012;**1**.

[4] Chen G, Yang C, Prasad PN. Nanophotonics and nanochemistry: controlling the excitation dynamics for frequency up-and down-conversion in lanthanide-doped nanoparticles. Accounts of chemical research. 2013;**46**: 1474-1486.

[5] Saleh BE, Teich MC. Fundamentals of Photonics. John Wiley & Sons. 1991;**84**.

[6] Prasad PN. Nanophotonics. John Wiley & Sons. 2004;**21**.

[7] Prasad PN. Introduction to biophotonics. John Wiley & Sons. 2003;**8**.

[8] Courjon D. Near-field microscopy and near-field optics. World Scientific Publishing Company; 2003;**12**.

[9] Wallis RF, Tamir T, Stegeman GI, editors. Electromagnetic Surface Excitations: Proceedings of an International Summer School at the Ettore Majorana Centre, Erice, Italy, July 1–13, 1985. Springer Science & Business Media. 2013;**12**.

[10] Fillard JP. Near field optics and nanoscopy. World Scientific. 1996.

[11] Saiki T, Narita Y. Recent advances in near-field scanning optical microscopy. JSAP international. 2002;**5**:22-29.

[12] Takagahara T. Effects of dielectric confinement and electron-hole exchange interaction on excitonic states in semiconductor quantum dots. Physical Review B. 1993;**47**:4569.

[13] Wang Y, Herron N. Nanometer-sized semiconductor clusters: materials synthesis, quantum size effects, and photophysical properties. The Journal of Physical Chemistry. 1991;**95**:525-532.

[14] Kobayashi T. J-aggregates. World Scientific. 2012.

[15] Kahnert FM. Numerical methods in electromagnetic scattering theory. Journal of Quantitative Spectroscopy and Radiative Transfer. 2003;**79**: 775-824.

[16] Miller EK. A selective survey of computational electromagnetics. IEEE Transactions on Antennas and Propagation. 1988;**36**:1281-1305.

[17] Myroshnychenko V, Rodríguez-Fernández J, Pastoriza-Santos I, Funston AM, Novo C, Mulvaney P, Liz-Marzán LM, De Abajo FJ. Modelling the optical response of gold nanoparticles. Chemical Society Reviews. 2008;**37**:1792-1805.

[18] Vandenbosch GA. Computational Electromagnetics in Plasmonics. Plasmonic-Principles and Applications. 2012;**24**:23-48.

[19] Smajic J, Hafner C, Raguin L, Tavzarashvili K, Mishrikey M. Comparison of numerical methods for the analysis of plasmonic structures. Journal of Computational and Theoretical Nanoscience. 2009;**6**: 763-774.

[20] Valev VK, Clercq BD, Zheng X, Denkova D, Osley EJ, Vandendriessche S, Silhanek AV,

Volskiy V, Warburton PA, Vandenbosch GA, Ameloot M. The role of chiral local field enhancements below the resolution limit of Second Harmonic Generation microscopy. Optics express. 2012; **20**:256-264.

[21] Goncalves MR. Plasmonic nanoparticles: fabrication, simulation and experiments. Journal of Physics D: Applied Physics. 2014;**47**:213001.

[22] Han K, Chang CH. Numerical modeling of sub-wavelength anti-reflective structures for solar module applications. Nanomaterials. 2014;**4**: 87-128.

[23] Gallinet B, Butet J, Martin OJ. Numerical methods for nanophotonics: standard problems and future challenges. Laser & Photonics Reviews. 2015;**9**:577-603.

[24] Montgomery JM, Lee TW, Gray SK. Theory and modeling of light interactions with metallic nanostructures. Journal of Physics: Condensed Matter. 2008;**20**:323201.

[25] Taflove A, Oskooi A, Johnson SG, editors. Advances in FDTD computational electrodynamics: photonics and nanotechnology. Artech house. 2013.

[26] Yee K. Numerical solution of initial boundary value problems involving Maxwell's equations in isotropic media. IEEE Transactions on antennas and propagation. 1966;**14**:302-307.

[27] Jin JM, Riley DJ. Finite element analysis of antennas and arrays. John Wiley & Sons. 2009;**23**.

[28] Busch K, Koenig M, Niegemann J. Discontinuous Galerkin methods in nanophotonics. Laser & Photonics Reviews. 2011;**5**:773-809.

[29] Niegemann J, Pernice W, Busch K. Simulation of optical resonators using DGTD and FDTD. Journal of Optics A: Pure and Applied Optics. 2009;**11**:114015.

[30] Martin OJ, Piller NB. Electromagnetic scattering in polarizable backgrounds. Physical Review E. 1998;**58**:3909.

[31] Draine BT, Flatau PJ. Discrete-dipole approximation for scattering calculations. Josa a. 1994;**11**:1491-1499.

[32] De Abajo FG, Howie A. Retarded field calculation of electron energy loss in inhomogeneous dielectrics. Physical Review B. 2002;**65**:115418.

[33] Kern AM, Martin OJ. Surface integral formulation for 3D simulations of plasmonic and high permittivity nanostructures. JOSA A. 2009;**26**: 732-740.

[34] Taboada JM, Rivero J, Obelleiro F, Araújo MG, Landesa L. Method-of-moments formulation for the analysis of plasmonic nano-optical antennas. JOSA A. 2011;**28**:1341-1348.

[35] Weiland T. A discretization method for the solution of Maxwell's equations for six-component fields.-Electronics and Communication, Archiv für Elektronik und Uebertragungstechnik. 1977:116-120.

[36] Mishchenko MI, Travis LD, Mackowski DW. T-matrix computations of light scattering by nonspherical particles: A review. Journal of Quantitative Spectroscopy and Radiative Transfer. 1996:535-575.

[37] Boyack R, Le Ru EC. Investigation of particle shape and size effects in SERS using T-matrix calculations. Physical Chemistry Chemical Physics. 2009;**11**: 7398-7405.

[38] Khlebtsov NG. T-matrix method in plasmonics: An overview. Journal of Quantitative Spectroscopy and Radiative Transfer. 2013;**123**:184-217.

[39] Hafner C. The generalized multipole technique for computational electromagnetics. Artech; 1990.

[40] Hafner C. Post-modern electromagnetics. Wiley. 1998.

[41] Baffou G, Quidant R, García de Abajo FJ. Nanoscale control of optical heating in complex plasmonic systems. ACS nano. 2010;**4**:709-716.

[42] De Abajo FG. Multiple scattering of radiation in clusters of dielectrics. Physical review B. 1999;**60**:6086.

[43] Moharam MG, Grann EB, Pommet DA, Gaylord TK. Formulation for stable and efficient implementation of the rigorous coupled-wave analysis of binary gratings. JOSA a. 1995;**12**: 1068-1076.

[44] Lalanne P, Jurek MP. Computation of the near-field pattern with the coupled-wave method for transverse magnetic polarization. Journal of modern optics. 1998;**45**:1357-1374.

[45] Franken EP, Hill AE, Peters CW, Weinreich G. Generation of optical harmonics. Physical Review Letters. 1961;7:118.

[46] Maiman TH. Optical and microwave-optical experiments in ruby. Physical review letters. 1960;4:564.

[47] Lewis GN, Lipkin D, Magel TT. Reversible photochemical processes in rigid media. A study of the phosphorescent state. Journal of the American Chemical Society. 1941;63: 3005-3018.

[48] Berini P, De Leon I. Surface plasmon–polariton amplifiers and lasers. Nature photonics. 2012;6:16-24.

[49] Stockman MI. The spaser as a nanoscale quantum generator and ultrafast amplifier. Journal of Optics. 2010;12:024004.

[50] Krasnok A, Alù A. Active nanophotonics. Proceedings of the IEEE. 2020;108:628-654.

[51] Boardman AD, Grimalsky VV, Kivshar YS, Koshevaya SV, Lapine M, Litchinitser NM, Malnev VN, Noginov M, Rapoport YG, Shalaev VM. Active and tunable metamaterials. Laser & Photonics Reviews. 2011;5:287-307.

[52] Zhang Y, Manjavacas A, Hogan NJ, Zhou L, Ayala-Orozco C, Dong L, Day JK, Nordlander P, Halas NJ. Toward surface plasmon-enhanced optical parametric amplification (SPOPA) with engineered nanoparticles: a nanoscale tunable infrared source. Nano letters. 2016;16:3373-3378.

[53] Kauranen M, Zayats AV. Nonlinear plasmonics. Nature photonics. 2012;6: 737-748.

[54] Zayats AV, Smolyaninov II, Maradudin AA. Nano-optics of surface plasmon polaritons. Physics reports. 2005;408:131-314.

[55] Uskov AV, O'Reilly EP, Manning RJ, Webb RP, Cotter D, Laemmlin M, Ledentsov NN, Bimberg D. On ultrafast optical switching based on quantum-dot semiconductor optical amplifiers in nonlinear interferometers. IEEE Photonics Technology Letters. 2004;16: 1265-1267.

[56] Kuntz M, Fiol G, Szewc C, Lammlin M, Meuer C, Bimberg D, Kovsh A, Ledentsov N, Ferber S, Schubert C, Jacob A. High-speed quantum dot lasers and amplifiers for optical data communication. Sixth IEEE Conference on Nanotechnology 2006; 17:702-705.

Section 3

Nonlinear Optical Properties of Novel Materials

Chapter 3

Green Synthesis of Metal Nanostructures and Its Nonlinear Optical Properties

Emusani Ramya

Abstract

Simple green synthesis of metal nanoparticles (Ag NPs) was prepared by using Raphanussativus leaf extract. This extract acts as reduce and stabilizing agent. The formation of silver NPs was confirmed and characterized by XRD, UV–visible absorption spectrum, TEM, and FTIR. The luminescence enhancement and quenching of $Eu^{3+}$and Sm^{3+} ions were observed in the presence of silver NPs. The luminescence enhancement is owing to arise in the electric-dipole transition with alteration of the field around Ln^{3+} ions. Nonlinear studies in femtosecond (fs) and picosecond (ps) time scales have been studied by using the Z-scan technique. Third-order nonlinear optical susceptibility of silver nanoparticles was obtained with Degenerate Four-Wave Mixing (DFWM) in the fs regime. The lifetimes of lanthanum complexes were increased by the concentration of silver NPs and decreased for further silver. The high enhanced luminescence and nonlinear studies of green synthesized silver nanoparticles can be used in optics and bio applications.

Keywords: Metal Nanostructures, Green synthesis, Luminescence enhancement, Nonlinear optical properties

1. Introduction

Nanoparticles are the building blocks for many materials and contribute to the rapid growth of nanoscience [1]. There has been wide interest in silver nanoparticles (Ag NPs) due to their unique properties arising from shape, size and composition finding their applications in sensing devices, bio-labelling, catalyst, electronics, photonics, surface-enhanced Raman spectroscopy (SERS) and biomedicine, etc. [2–7]. For the applications of environmental friendly NPs in various fields, development of easy preparation methods using nontoxic and nonhazardous materials and methods is needed. Biosynthesis route has many advantages compared to other synthesis methods, as these routes do not use high temperature, pressure, energy and toxic elements. Recently, several groups reported on the synthesis of biocompatible Ag NPs using natural sources like fungi, yeast, bacteria [8–12] and plant extracts [13–16]. The properties of biosynthesized metal or bimetallic.

NPs are different from those synthesized from chemical methods and laser ablation methods [17, 18] because they are highly stable, nontoxic and biocompatible as the surfaces of the NPs are coated with biogenic surfactants. Biosynthesized metal NPs have biomedical applications like antimicrobial coatings (e.g.

antibacterial, antifungal, antiviral activity, antiparasite) [19–22], drug deliveries [23], medical imaging [24], anticancer NPs [25], medical diagnostics, sensors [26], catalytic degradation of organic pollutants to enhance membrane treatment processes [27] and also as optical limiters [28, 29] in the field of optics.

The *Raphanussativus* leaves belong to the Brassicaceae family, call it Radish and it is root vegetable. Derivatives and various components of *Raphanussativus* consist of 2- Hexen −1- al (leaf aldehyde) 3- hexane −1- ol (leaf alcohol) and n- isobutyr-aldehyde and isovaleraldehyde, sapogenin, methin, levon, phosphatase, histamines and spasmolytic components, lysine, polyphenolic, Sulphoraphene, vanillic acids and raphanin, those are having various medicine utilities. These bio-molecules consist in the extract are binding on the surface of nanoparticles and reduce the ions to NPs, additionally keeping stabilize the nanoparticles.

Lanthanide (Ln^{3+}) complexes are of great interest due to their large Stoke shifts, long emission lifetimes, and narrow emission bandwidths. Metal-enhanced luminescence of rare-earth ions is of great interest because of their applications in laser materials, nano-photonics, LEDs, and biosensors [30]. Few groups investigated that enhancement or quenching of luminescence is depending on the concentration, shape, and size of metal nanoparticles based on the interaction of metal nanoparticles and luminescence centres [31, 32]. In the presence of metallic nanostructures, the field around the rare-earth ions can alter the emission and excitation [33, 34]. The literature reported that luminescence enhancement is because of energy transfer between Ln^{3+} complexes and metal NPs [35]. Quantum yields and lifetimes are influenced by changing the decay rate of radiative and non-radiative transitions [36]. Nevertheless, the process elaborated in Ln^{3+} ion luminescence with Ag NPS is not clear and hence, essential to address which is a promising and fascinating challenge.

Here, we have synthesized silver nanoparticles by using *Raphanussativus* leaf extract. The aim of the work is to discuss the origin of the effect of Ag NPs on enhancing and quenching the luminescence intensity of Ln^{3+} complexes and their nonlinear optical properties.

2. Experimental details

2.1 Materials and methods

$AgNO_3$ (Silver nitrate, 99.99%) was procured from Sigma Aldrich. *Raphanussativus* leaves were got from the market. Lanthanum complexes Eu $(TTFA)_3$ and $Sm(TTFA)_3$ were obtained from Soltech BioScience and Rare Earth products Inc., USA.

2.2 Preparation of Raphanussativus leaf extract

First Raphanussativus leaves were wiped using distilled water. 10 grams of *Raphanussativus* leaves have taken into 100 ml of distilled water and heated for 5 minutes and grinded in a pestle. The mixture solution was filtered with Whatman no. 1 filter twice to get rid of granules and refrigerated at 4°C.

2.3 Synthesis of Ag NPs using Raphanussativus leaf extract

1.5 ml of freshly prepared leaf extract was poured into 30 ml of silver nitrate (1 mM) aqueous solution with stirring at room temperature. Then, the solution changed from transparent to ash color in 5 minutes, which represents the formation

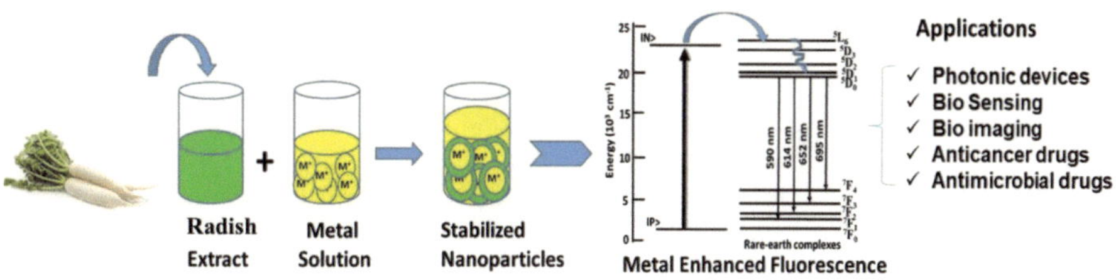

Figure 1.
Synthesis of Ag nanoparticles.

of silver nanoparticles. The nanoparticles were stable for few months. Preparation process can be shown in **Figure 1**.

2.4 Sample preparation for luminescence studies

Rareearth ions europium and samarium of weight ratio of 0.28% were dissolved. Various concentrations of silver nanoparticles (0.16, 0.33, 1, 1.6, 2.32, 3.32 and 6.64 μM) were added to 0.10, 0.13, 0.20 μM of europium and 1.32, 1.45, 1.58 μM of samarium for luminescence studies.

3. Characterization techniques

The green synthesized silver nanoparticles and Lanthanide complexes with silver nanoparticles were characterized using XRD, TEM, FT-IR, UV–visible absorption, luminescence studies, Z-scan, degenerate four-wave mixing (DFWM), and decay measurements techniques. XRD pattern of silver nanoparticles was measured on Cu-K_α X-ray radiation with $\lambda = 1.5406$ Å over the range of $2\theta = 5$–$50°$ at 1° min^{-1}. A transmission electron microscope (TEM) on FEI TECHNAI G^2 S-Twin was used to measure the diameter and morphology of nanoparticles. FT-IR spectrum was performed using Thermo-Nicolet 6700 spectrophotometer with a 500–4000 cm^{-1} range. UV–Vis spectra were carried out by JASCO-V670 UV/VIS/NIR spectrometer with 1 nm resolution and 200–800 nm range. Luminescence emission spectra of Ln^{3+} complexes with silver nanoparticles were performed on HORIBA JOBIN YVON spectrophotometer with 262 nm and 350 nm excitations. Nonlinear absorption and optical limiting studies were performed using the Z-scan with laser source of Nd: YAG laser at 532 nm, 30 ps and 10 Hz repetition rate; and 1 kHz, 110 fs pulses of Ti: sapphire at wavelength 800 nm. Degenerate four-wave mixing (DFWM) measurements were studied with Ti: sapphire laser in box-car geometry. Luminescence decay studies were performed on FLS980 with an excitation source as a Xenon lamp.

4. Results and discussions

4.1 X-ray diffraction (XRD)

XRD spectrum of Ag NPs confirms the crystalline nature in **Figure 2**. Diffraction patterns of silver nanoparticles at 38.1°, 43.9°, 64.6° and 77.3° and corresponding the planes (111), (200), (220) and (311) are compared with standard spectrum [37]. Which showed the silver nanoparticles are having FCC formation. The diffraction reflection at (111) is sharp and high intensity as compared to the planes at (200),

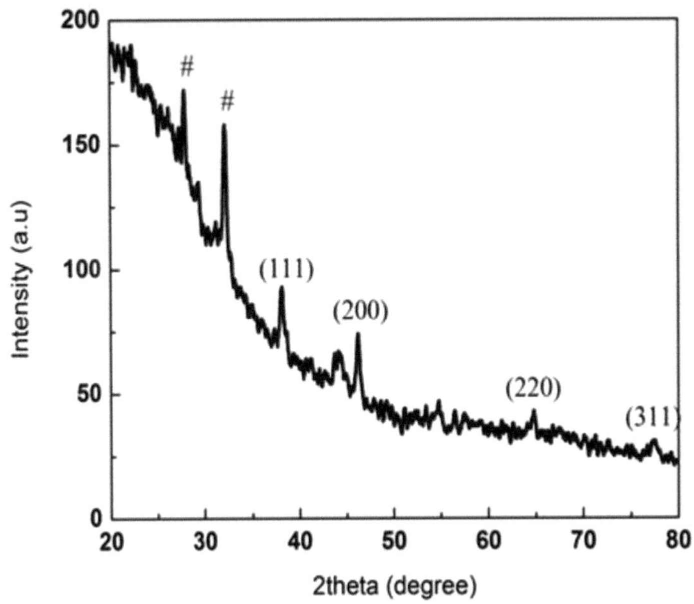

Figure 2.
X-ray diffraction spectrum of silver nanoparticles.

(220), and (311). It demonstrates that silver nanocrystals were oriented along the (111) plane. The unassigned reflections at marked with #, 27.7° and 32.2° were related to the biological phase appearing on the silver nanoparticles' surface. Therefore, these results indicate that the silver nanoparticles were formed. Both the TEM results and XRD results are in good agreement.

4.2 Transmission electron microscope (TEM)

The morphology, size distribution, and selected area electron diffraction (SAED) circular patterns of silver nanoparticles were exhibited in **Figure 2**. From **Figure 3(a–c)**, the Ag NPs are nanospheres in shape, crystalline, and size is about 7 nm respectively. The planes (111), (200), (220), and (311) from the SAED pattern were exhibiting FCC structure from **Figure 3(d)**.

4.3 UV-Vis absorption spectroscopy

UV–Vis linear absorption of silver nanoparticles with 240–800 nm range areas displayed in **Figure 4**. The absorption spectra were measured at various concentrations of silver nanoparticles added to 3 ml of distilled water and silver nanoparticles with Eu and Sm complexes at various concentrations.

UV–visible absorption spectrum of various concentrations are (0.16, 0.33, 1.0, 1.60, 2.32, 3.32, 6.64 and 9.96 μM) of silver nanoparticles as depicted in **Figure 3(a)**. The band at 262 nm is becoming strong as increase the concentration because the number of organic compounds absorption is increasing, which helps to reduce the silver ions. The plasmonic band of silver nanoparticles at 460 nm is increasing with an increase in the concentration from 0.16 μM to 9.96 μM. The inset shows the absorption of silver NPs is increasing linearly at 460 nm i.e. the nanoparticles are not agglomerating with concentration.

The absorption spectrum of europium complex with various concentrations (0.13, 0.20, 0.26, 0.33, 0.99, 1.98 and 3.3 μM) as displayed in **Figure 4(b)**. The linear absorption peaks at 262 nm and 342 nm corresponding to $\pi \rightarrow \pi^*$ transition

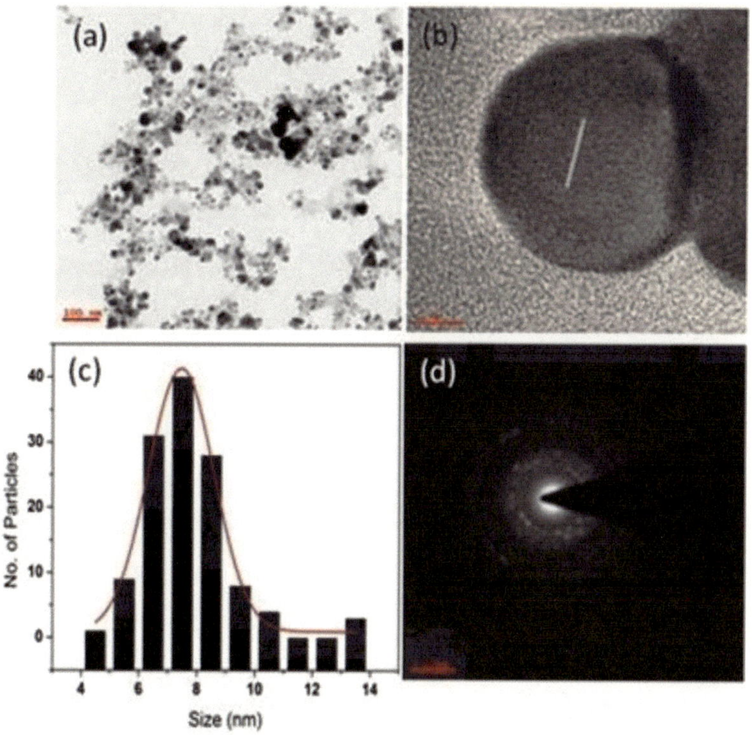

Figure 3.
(a) TEM figures (b) high-resolution TEM (HRTEM) image (c) particle size distribution and (d) SAED circular patterns of silver nanoparticles.

due to complex absorbance and n → π* transition of europium and occurrence of the shoulder at (1.99 and 3.3 μM) higher concentrations due to strong interaction among the europium molecules [38].

The absorption spectrum of europium complex (0.13 μM) with various concentrations (0.16, 0.33, 1.0, 1.60, 2.32, 3.32, and 6.64 μM) of silver nanoparticles (Ag NPs) and the absorption bands appeared at 262 nm, 342 nm, and 460 nm as seen in **Figure 4(c)**. The absorption band in the range of 300–400 nm is shifted to red slightly because the interaction of silver nanoparticles and europium molecules i.e. the Plasmon field of silver nanoparticles is influencing the europium ions but the surface plasmonic peak of Ag NPs at 460 nm is not altering. The inset shows a linear increase in absorption with silver NPs, suggesting that nanoparticles are not agglomerating in the complex solution.

Absorption spectrum of samarium complex (1.45 μM) with various concentrations (0.16, 0.33, 1.0, 1.60, 2.32 and 3.32 μM) of silver nanoparticles as depicted in **Figure 5**. The absorption band at 265 nm corresponds to the π → π* transition of the samarium complex and biological components in the extract, and bands at 342 nm and 460 nm correspond to the n → π* transition of samarium and plasmonic peak (SPR) of silver nanoparticles. Samarium complex concentration (1.45 μM) is more as compared with silver nanoparticles so the bands of samarium (265 nm and 342 nm) are predominating the SPR.

4.4 Fourier transform infrared (FTIR) technique

FT-IR spectrum displays the *Raphanussativus* leaf extract used for biosynthesis of Ag NPs in **Figure 6**. The Silver suspension was loaded on a potassium bromide (KBr) pellet and dried. The FT-IR peaks of silver nanoparticles appeared at

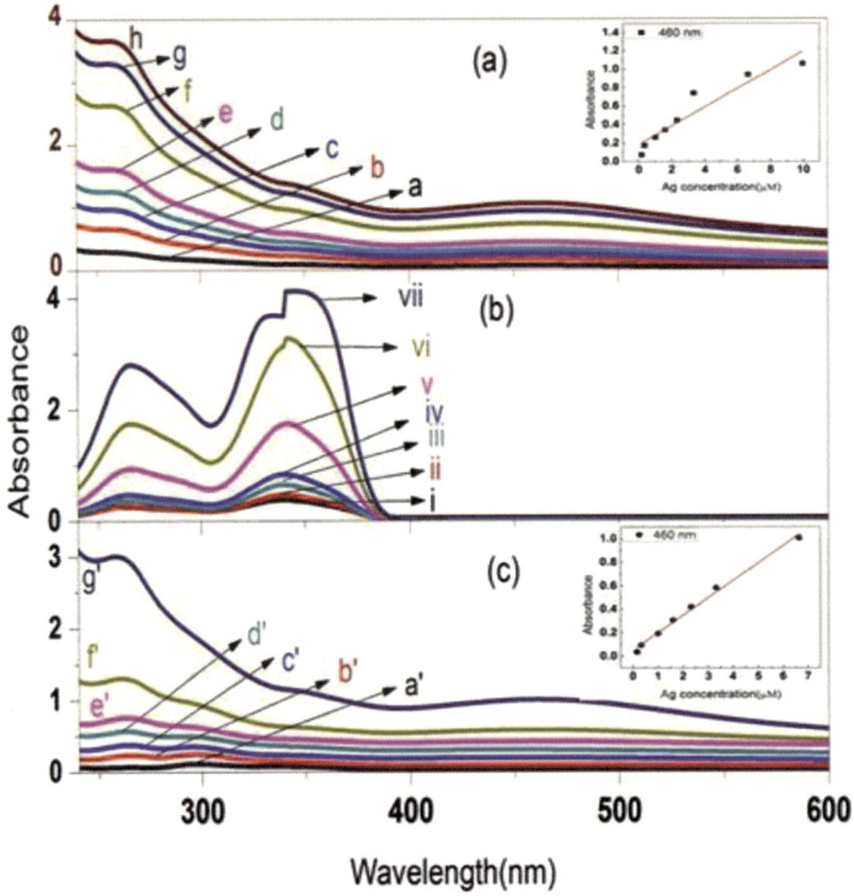

Figure 4.
UV–visible spectra of (a) silver nanoparticles with various concentrations: a) 0.16, b) 0.33, c) 1.0, d) 1.60, e) 2.32, f) 3.32, g) 6.64 and h) 9.96 µM. Inset displays the linear absorption of silver nanoparticles at 460 nm with increasing silver. (b) Various concentrations of europium ions in ethanol are: i) 0.13, ii) 0.20, iii) 0.26, iv) 0.33, v) 0.99, vi) 1.98 and vii) 3.3 µM. (c) Various concentrations of Ag NPs with 0.13 µM europium solution: A') 0.16, b') 0.33, c') 1.0, d') 1.60, e') 2.32, f') 3.32 and g') 6.64 µM. Inset figure exhibits the absorption of silver nanoparticles at different concentrations in (0.13 µM) europium solution at 460 nm.

3315 cm^{-1} indicate hydroxyl –OH stretching; 2929 cm^{-1} and 2834 cm^{-1} assigned to stretching C-H modes of methyl groups; 1636 cm^{-1} assigned to carbonyl (–C=C) stretching; 1382 cm^{-1} attributed to –C-O stretching mode of water-soluble organic components like polyphenols, alkaloids, and flavonoids in *Raphanussativus* extract; 1040 cm^{-1} indicate the C-O alcoholic stretching group. These results concluded that the organic compounds of the extract are responsible for making nanoparticles [39–42].

4.5 Photoluminescence studies

The excitation spectra of europium complex (λ_{em} = 614 nm) at various concentrations (0.03, 0.07, 0.13, 0.20, 0.26, 0.66, 1.99 and 3.3 µM) as shown in **Figure** 7. The bands in the range 250–400 nm are due to $\pi \rightarrow \pi^*$ transitions of Eu. At very low concentrations of europium (0.03 µM and 0.07 µM), the broadband appears at 340 nm. As increase the concentration (0.13–0.66 µM), the new band appeared at 270 nm and 340 nm band splits into two bands which results, strong interactions among the europium complex in the solution phase and depend on the

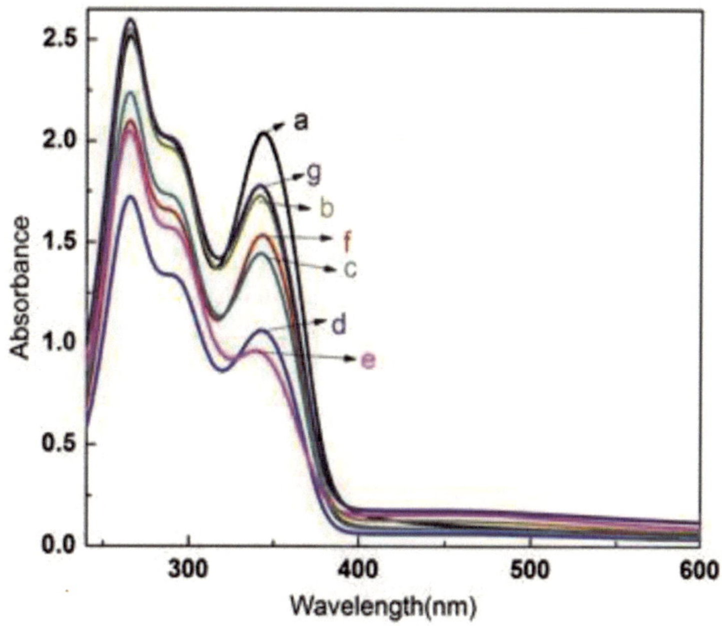

Figure 5.
Linear absorption spectrum of samarium complex (1.45 μM) with various concentrations are a) 0.166 b) 0.33 c) 1.0 d) 1.60 e) 2.32 f) 3.32 and g) 9.96 μM of Ag NPs.

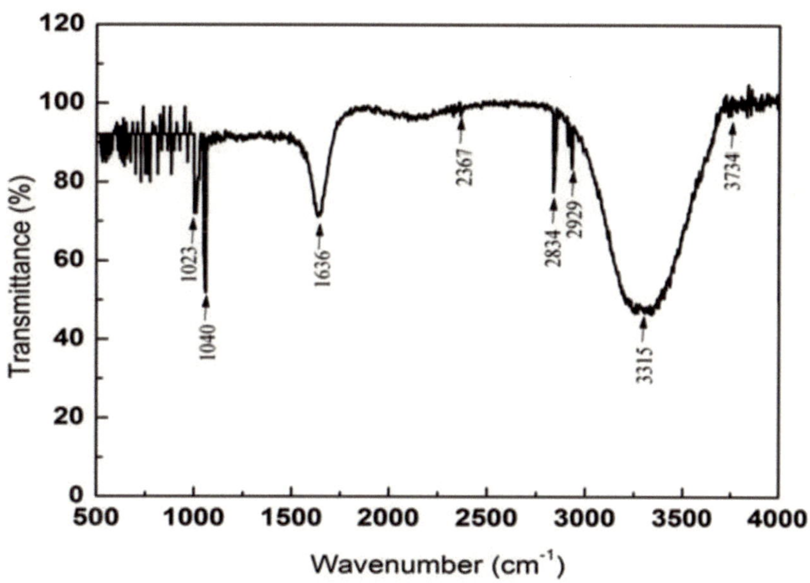

Figure 6.
FT-IR Spectrum of Raphanussativus *leaf extract used for biosynthesized silver nanoparticles.*

concentration of europium. As increase the concentration further (1.99 μM and 3.3 μM), 270 nm and 340 nm bands shifted to the blue region, and the 270 nm band has vanished. Even for further concentrations, the 370 nm band shifted to the red region, which indicates that the interaction among europium ions enhanced.

The emission spectra of the europium complex excited with 262 nm as shown in **Figure 8**. The band at 614 nm ($^5D_0 \rightarrow {}^7F_2$) is a hypersensitive electric-dipole

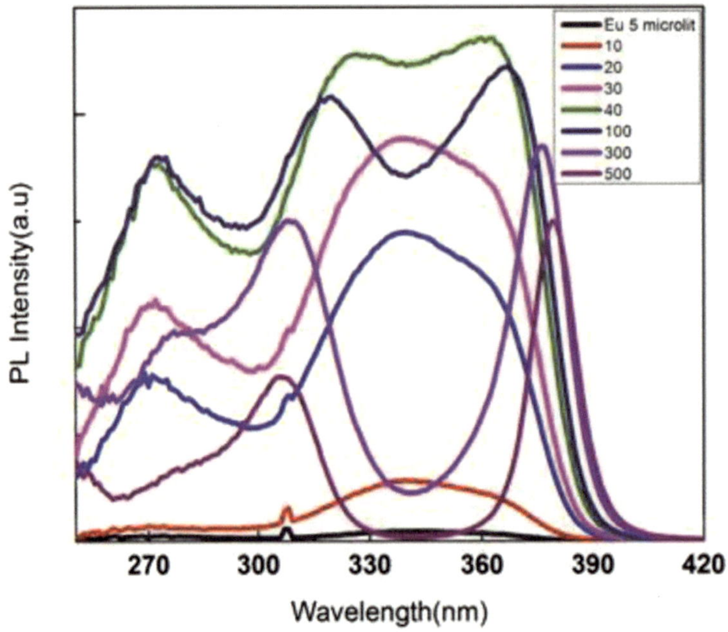

Figure 7.
Excitation spectra of europium complex (λ_{em} = 614 nm) at various concentrations (0.03, 0.07, 0.13, 0.20, 0.26, 0.66, 1.99 and 3.3 μM).

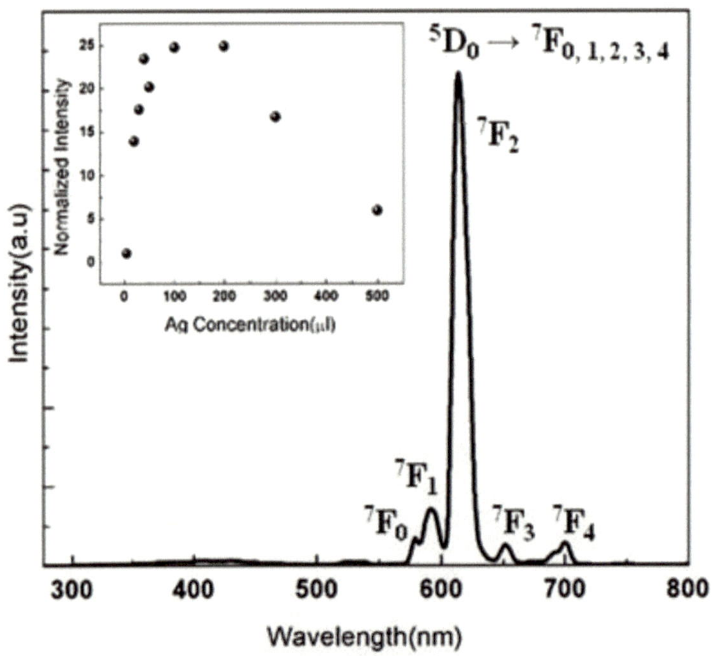

Figure 8.
Emission spectra of europium complex (λ_{exc} = 262 nm). The inset figure displays the $^5D_0 \rightarrow {}^7F_2$ transition on various silver nanoparticle in europium complex (0.13 μM).

transition, a dominant peak. The magnetic dipole transitions are at 577 nm and 590 nm ($^5D_0 \rightarrow {}^7F_0$ and $^5D_0 \rightarrow {}^7F_1$). Inset displays the figure of concentration of silver nanoparticles with the intensity of electric dipole transition ($^5D_0 \rightarrow {}^7F_2$) in europium complex. As increase, the concentration of silver nanoparticles, affect the ligand field surrounding the europium ions, consequently enhance the electric dipole transition rate. The $^5D_0 \rightarrow {}^7F_2$ transition enhanced its intensity 25 times up to the 6.64 μM of silver nanoparticle and quenching slowly exceeding 6.64 μM in 0.13 μM europium solution. Hence, we emphasize the effect of silver on the europium luminescence emission intensity of electric dipole transition ($^5D_0 \rightarrow {}^7F_2$).

The emission spectra of europium (λ_{exc} = 350 nm) with various (0.16, 0.33, 1.0, 1.60, 2.32, 3.32 and 6.64 μM) of silver nanoparticles as seen in **Figure 9**. The luminescence emission intensity of $^5D_0 \rightarrow {}^7F_2$ transition starts enhanced and maximum at 1.6 μM of silver NPs and gets quenched for further increase the silver in europium (0.13 μM) and the enhancement factor is ~5. In the spectra of excitation, Ag NPS cannot intensify the europium ions excitation at 262 nm and 350 nm, nanoparticles absorption (400–550 nm) is uncertain at excitation wavelengths, although it enormously affects the luminescence centers of europium ions in the emission. At excitation wavelength around absorption of silver (350 nm), the intensity of the electric field increased certain times induce an intensification in luminescence intensity by a few hundred times. The enhancement of luminescence intensity because of the overlap of europium emission and the scattering of nanoparticles which bank on the gap of NPS [33, 43, 44]. The distance of NPS reduces, the scattering at 612 nm overlay on the emission enlarges appearing in the increment of luminescence intensity. Similarly, the intensity decreases with the distance of nanoparticles increases. The luminescence increment was observed only at 0.13 μM concentration of europium and at other concentrations of 0.10, and 0.20 μM, the luminescence is quenching. That means, the enhancement of luminescence intensity purely depends on the concentration, distance, size, and shape of nanoparticles.

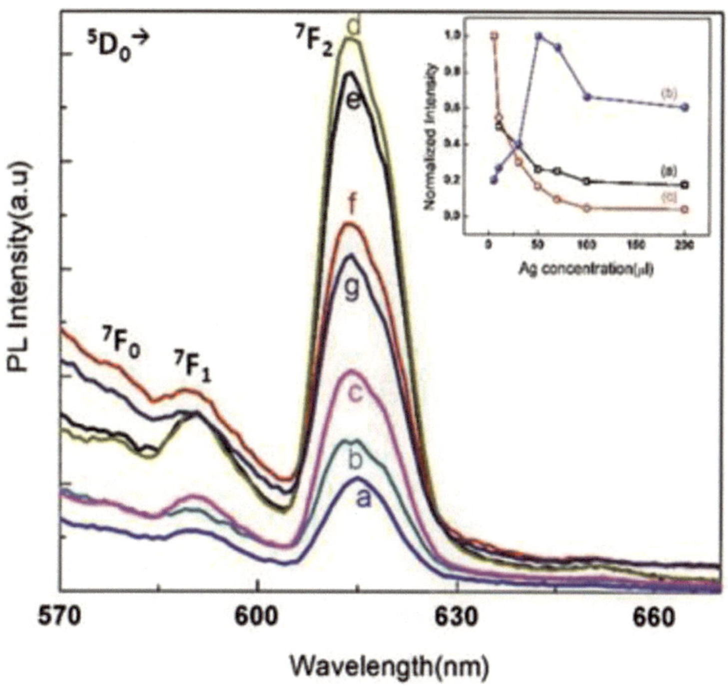

Figure 9.
Emission spectra of europium with various silver nanoparticles: a) 0.16, b) 0.33, c) 1.0, d) 1.60, e) 2.32, f) 3.32 and g) 6.64 μM excited with 350 nm. Inset figure shows dependence of luminesce intensity with silver concentration. Eu concentrations are (a) 0.1 (b) 0.13 and (c) 0.2 μM.

At 260 nm excitation, the enhancement factor is high as compared with 350 nm excitation, which is caused by the biological components in extract and ligand enhance the luminescence effectively and the overlap of emission europium and scattering of silver, luminescence efficiency will be high.

To understand the luminescence enhancement alteration on various lanthanides, we also obtained the emission of the samarium complex. The emission spectra of samarium with different concentrations of silver nanoparticles (0.16, 0.33, 1.0, 1.60, 2.32, 3.32, and 6.64 μM) at 350 nm excitation as can be seen in **Figure 10**. The inset picture exhibits the luminescence emission versus silver concentration with varying samarium concentrations (1.32, 1.45, and 1.58 μM). The luminescence intensity is noticed at 1.45 μM of samarium. The transitions at 645 nm ($^4G_{5/2} \rightarrow {}^6H_{9/2}$) electric dipole, 566 nm ($^4G_{5/2} \rightarrow {}^6H_{5/2}$), and 602 nm ($^4G_{5/2} \rightarrow {}^6H_{7/2}$) are magnetic dipole transitions. The enhancement factor of electric dipole transition is 7.4 at 2.32 μM of silver. For more increases in silver, the enhancement factor decreases [45, 46]. The enhancement factors for magnetic dipole transitions at 566 nm are 1.9 and 602 nm is 5.2. The scattering can alter by altering the distance of nanoparticles. Hence, the change in luminescence intensity overlay on the scattering and emission spectra as in the case of europium. So this is the reason for emission intensity quenched at 1.32 μM and 1.58 μM of samarium as depicted in the inset.

4.6 Nonlinear optical properties

Nonlinear absorption coefficients were obtained by using the Z-Scan technique with Nd: YAG laser, repetition rate 10 Hz, 30 ps pulses with 532 nm, and Ti: Sapphire laser, 1 kHz, 110 fs pulses with 800 nm. The nonlinear absorption studies of the biosynthesized silver nanoparticles solution were studied using an open aperture Z-scan set up (**Figure 11**) [29, 47]. In the Z-scan setup, the Gaussian

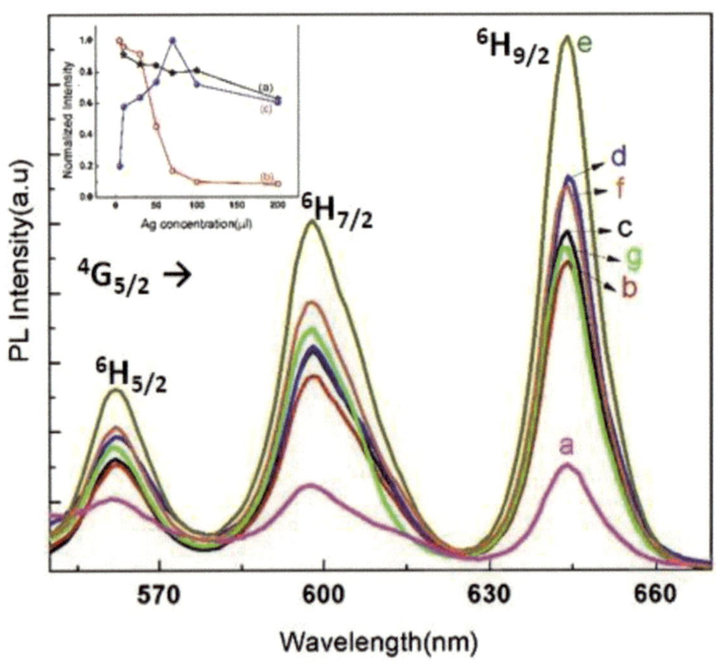

Figure 10.
Emission spectra of samarium 1.45 μM with silver nanoparticles: a) 0.16, b) 0.33, c) 1.0, d) 1.60, e) 2.32, f) 3.32 and g) 6.64 μM excited with 350 nm. Inset shows the dependence PL intensity on silver concentration at various samarium concentrations (a) 1.32 (b) 1.45 and (c) 1.58 μM.

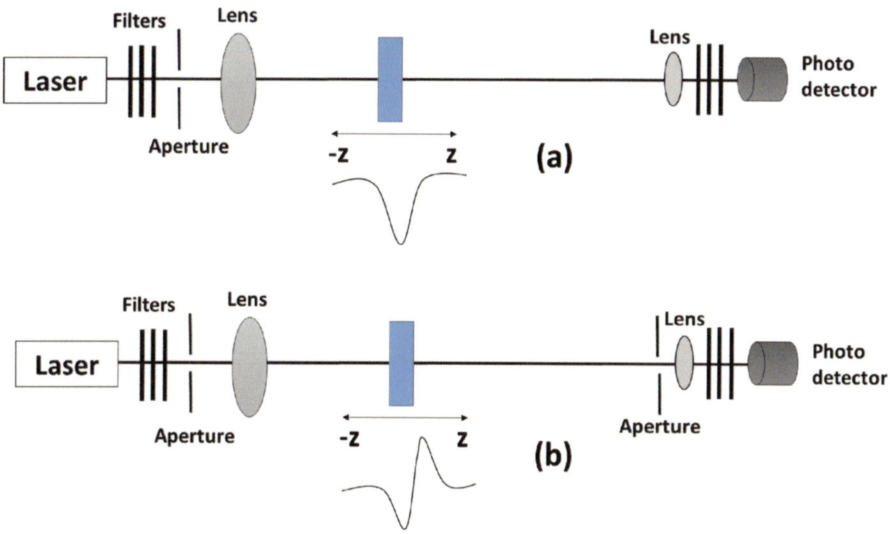

Figure 11.
Z-scan experimental set up a) open aperture b) closed aperture.

profile of the laser beam is concentrated with the lens. The silver solution in a 1 mm thickness quartz cuvette is moving along the Z-direction through-beam-focused direction. A focus point, the silver sample undergoes maximal intensity and slowly reduce from the focal point in both directions and the $f/40$ configuration is operated here. The width of the sample undergoes less than the Rayleigh range (3 mm). For beam, shaping apertures are used, and to change the intensity of laser neutral density filters are used. The experimental data is measured by examining the sample along with the focus and saving the data by boxcar averager (model SRS 250) with an analog-to-digital (ADC) card to a computer. The absorption coefficient (α_2) for open aperture Z scan is measured by fitting the transmittance equation

$$T_{OA(2PA)} = 1 - \frac{\alpha_2 I_0 L_{eff}}{2^{3/2}\left(1 + z^2/z_0^2\right)} \tag{1}$$

Where, I_0 - intensity at focus on the sample, z - sample position, λ - laser wavelength, $z_0 = \pi\omega_0^2/\lambda$ is Rayleigh range, ω_0 - beam waist at the focus (Z = 0), α_2- nonlinear absorption coefficient, and L_{eff} - effective path length is $L_{eff} = \frac{1\text{-}e^{-\alpha_0 L}}{\alpha_0}$, L - sample length, α_0 is the linear absorption coefficient.

Figure 12(a) displays the Z-scan data of open aperture and varying input intensities of silver nanoparticles. The symbols are the experimental data, those are fitting theoretically (solid curves) using Eq. 1. These curves showed reverse saturable absorption (RSA) behavior in biosynthesized silver nanoparticles, which are ascribed to the excitations from the plasmonic band to the free carrier absorption band of silver nanoparticles and two-photon absorption (TPA) from the ground state. We determined α_2 is 10.2×10^{-9} cm^2/W at various intensities 1.2 GW/cm^2–3.9 GW/cm^2. These coefficients are not altering much at input intensities.

Optical limiting experimental curves of silver nanoparticles as shown in **Figure 12(b)**. The threshold optical limiting value is 4 mJ/cm^2. Nonlinear scattering is not observed so optical limiting is due to two-photon absorption (TPA) and excited-state absorption (ESA) from SPR.

Similarly, **Figure 13(a)** illustrates the open aperture Z-scan data measured for silver nanoparticles with a wavelength of 800 nm at different input intensities from

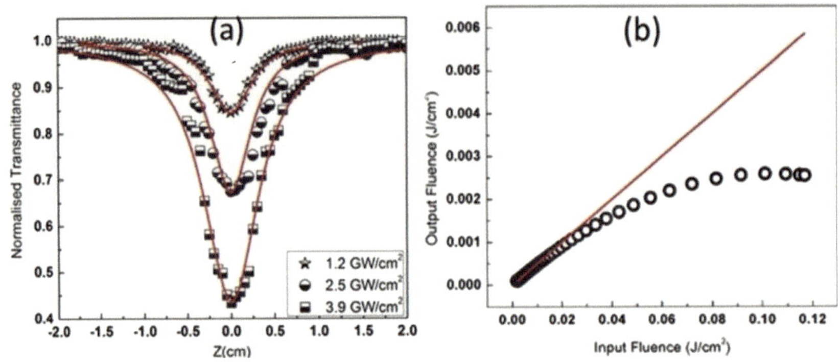

Figure 12.
(a) Z-scan of silver nanoparticles with various input intensities and (b) optical limiting curve with 532 nm.

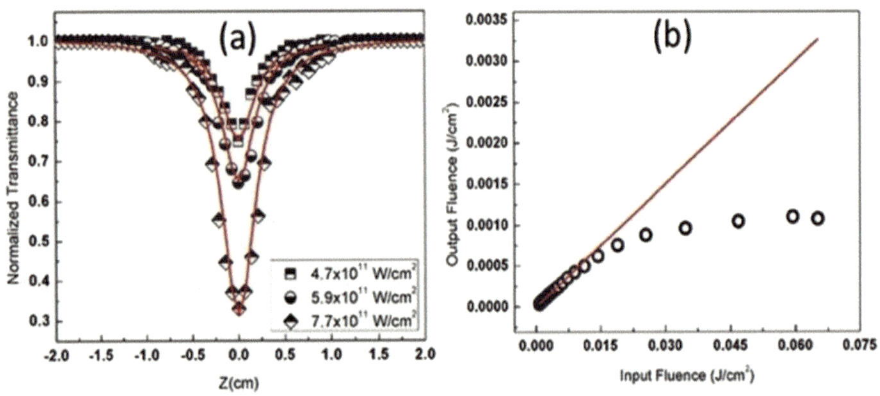

Figure 13.
(a) Z-scan of silver nanoparticles with various input intensities and (b) optical limiting curve with 800 nm.

4.7×10^{11}–7.7×10^{11} W/cm^2. Solid lines give the fits theoretically acquired by Eq. (2). RSA behavior of open aperture data gives the two-photon absorption (TPA). The two-photon absorption coefficients are $1.6 \times 10^{-9} - 3.8 \times 10^{-9}$ cm^2/W measured from theoretical fitting. The optical limiting data of silver nanoparticles with femtosecond laser is shown in **Figure 13(b)**. The optical limiting threshold value is 1.2 mJ/cm^2. So bio-reduced silver nanoparticles displaying good optical limiting behavior in both regimes and silver nanoparticles can behave as broadband optical limiters [29, 48].

4.7 Degenerate four-wave mixing technique (DFWM)

By using Ti: Sapphire laser with 800 nm, the third-order nonlinear susceptibility was obtained. Carbon disulfide (CS_2) has taken reference to the same input powers to measure third-order nonlinear susceptibility $\chi^{(3)}$ of silver nanoparticles using Degenerate Four Wave Mixing set up (**Figure 14**). The temporal profile of biosynthesized silver nanoparticles is shown in **Figure 15**. The cubit fit of the DFWM signal provides the nature of the third-order susceptibility. The inset picture gives the slope of silver nanoparticles, it is approximately $\sim$4. It shows that the DFWM signal has the contribution of two-photon absorption due to the electronic polarizability of the ground state alone. The $\chi^{(3)}$ was obtained as 2.95×10^{-14} esu for silver nanoparticles by the following equation

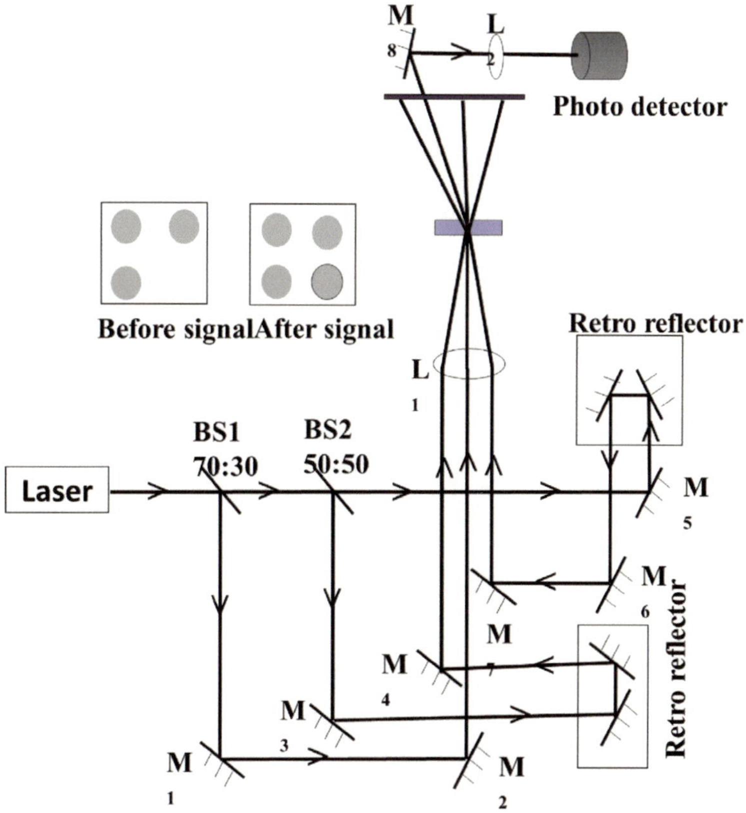

Figure 14.
Four wave mixing experimental set up.

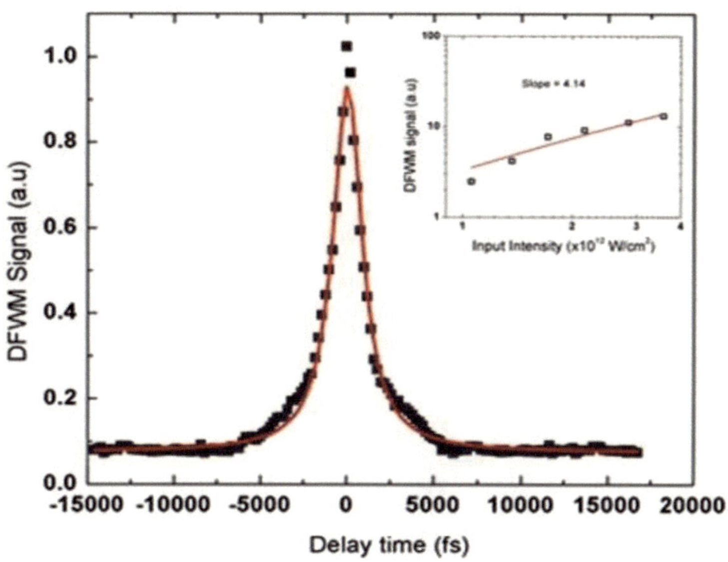

Figure 15.
Temporal response of DFWM signal of silver nanoparticles with 0.9 mg/l. inset picture gives the DFWM signal versus input intensity at zero delays. The solid line gives linear fit.

$$\chi^{(3)}_{sample} = \left(\frac{n_{sample}}{n_{ref}}\right)^2 \left(\frac{I_{sample}}{I_{ref}}\right)^{1/2} \left(\frac{L_{ref}}{L_{sample}}\right) \alpha L_{sample} \left(\frac{e^{\frac{\alpha L_{sample}}{2}}}{1 - e^{-\alpha L_{sample}}}\right) \chi^{(3)}_{ref} \tag{2}$$

Where L - path length of the sample, n - refractive index, I - DFWM signal intensity, and α - absorption coefficient.

The second-order hyperpolarizability (γ) is measured by the equation.

$$\chi^{(3)}_{sample} = T^4 \left[N_{solvent}\gamma_{solvent} + N_{sample}\gamma_{sample}\right] \tag{3}$$

T - local field factor, $T = \frac{n^2+2}{3.}$

Where, n - refractive index, N- number density of the solvent and the sample, and γ is the second-order hyperpolarizability. The Number density N can be written as $N = N_0 C_s/1000$

Where N_0-Avogadro number and C_s -concentration of the solution

Second-order hyperpolarizability is obtained as 2.1 X10^{-32} esu for silver nanoparticles. The nonlinear refractive index is measured from $\chi^{(3)}$ as 6.57 x 10^{-16} cm^2/W. For reproducibility, the experiment was repeated twicely [49].

4.8 Decay measurements

The decay curve of europium ions ($^5D_0$ level) in various concentrations of silver nanoparticles was obtained with 350 nm excitation. The decay profile of 0.13 μM of europium ions with silver nanoparticles (1.60 μM) was measured by monitoring the $^5D_0 \rightarrow {}^7F_2$ transition at 612 nm as shown in **Figure 16(a)**. The decay curves are having single exponential behavior. Average decay time (τ) is obtained by below equation

$$\tau = \frac{\int tI(t)dt}{\int I(t)dt} \tag{4}$$

From the data, it is clear that the lifetime increase with the concentration of Ag NPs (1.60 μM) then decreases rapidly for further increase the silver at a particular concentration of europium. The lifetime was increased from 275 μs to 361 μs, from 0 to 1.60 μM concentration of Ag NPs, then decreased lifetime for further increase the nanoparticles. The alteration of a lifetime follows the same tendency as the emission

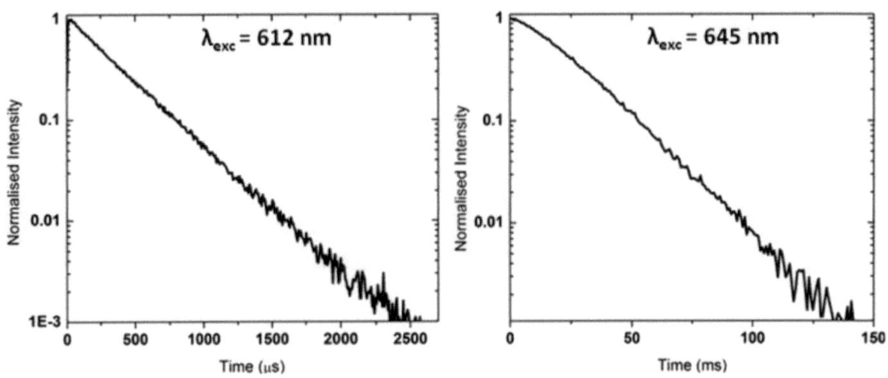

Figure 16.
Fluorescence decay rate of (a) europium and (b) samarium complexes monitoring at 612 nm and at 645 nm bands excited with 350 nm in 1.60 and 2.32 μM of silver nanoparticles respectively.

intensity in the presence of Ag. As the increase in the asymmetric ratio ($^5D_0 \rightarrow {}^7F_2$/ $^5D_0 \rightarrow {}^7F_1$), the decay rate of radiative transition increases than the decay rate of non-radiative transition of europium [50]. This shows that the presence of nanoparticles reduces the loss of energy due to non-radiative decay [51].

Similarly, the decay profile of the $^4G_{5/2}$ level samarium (1.45 μM) ions in the presence of silver NPs (2.32 μM) excited at 350 nm and monitoring 645 nm emission band as shown in **Figure 16(b)**. The decay is also single exponential with silver. Average decay τ values for samarium increases from 16 μs to 31 μs for 0 to 2.32 μM of silver and then reduces with an additional increase of silver. This behavior is due to enhancing in radiative decay rate. The samarium ions transmit energy to nanoparticles, then excited SPR converted into emission with enhancing luminescence efficiency consequently increase in lifetime. This is assigned to the luminescence emission intensity enhancing and quenching exhibited by samarium ions with silver. The lifetime of samarium ions is smaller as compared with europium ions may be because higher concentrations reduce the nonradiative decay rates.

5. Conclusions

- Introduced fast, simple, economic, and eco-friendly biological synthesized silver nanoparticles using *Raphanussativus* leaf extract.

- The Luminescence enhancement factor for europium ions is high compared to samarium ions. Rare-earth ions displaying luminescence emission intensity enhancement and quenching with silver nanoparticles are assigned to local field enhancement and decay processes.

- Third-order nonlinear optical properties of biosynthesized silver nanoparticles were studied in femtosecond and picosecond regimes.

- Nonlinear susceptibility was measured by the DFWM technique.

- Ag NPs show good nonlinear optical absorption and optical limiting properties.

- These non-toxic NPs show many applications such as luminophores in biomedical and the strong reverse saturable absorption make them be good optical limiters.

Author details

Emusani Ramya
Department of Science and Humanities, CMR Institute of Technology, Hyderabad, Telangana, India

*Address all correspondence to: eramya29@gmail.com

References

[1] Geddes C D, Lakowicz J R (2002) Metal-Enhanced Fluorescence. *J Fluoresc*. 12:121-128

[2] Frederix F, Friedt J M, Choi K H, Laureyen W, Campitelli A, Mondelaers D, Maes G, Borghs G (2003) Biosensing based on light absorption of nanoscaled gold and silver particles. *Anal Chem* **75**: 6894-6900

[3] Wang J (2003) Nano particle-based electrochemical DNA detection. *Analytica Chimica Acta*. 500: 247-257

[4] Tripathi G N R (2003) P-Benzosemiquinone Radical Anion on Silver nanoparticles in water. *J Am Chem Soc* 125:1178-1179

[5] Tian Z Q, Ren B (2004) Adsorption and reaction at electrochemical interfaces as probed by surface-enhanced Raman spectroscopy. *Annu Rev Phys Chem* 55:197-229

[6] Parak W J, Gerion D, Pellegrino T, Zanchct D, Micheel C, Williams S C, Boudreau R, Le Gros M A, and Larabell C A, Alivisatos A P (2003) Biological Applications of Colloidal Nanocrystals. *Nanotech* 14:15-27

[7] Mansoori G A, George T F, Zhang G, Assoufid L (**2007**) Molecular building blocks for nanotechnology. Springer, New York

[8] Nair B, Pradeep T (2002) Coalescence of Nanoclusters and Formation of Submicron Crystallites Assisted by Lactobacillus Strains. *Cryst Growth Des* 2:293-298

[9] Klaus T, Joerger R, Olsson E, Granqvist C G (2001) Bacteria as workers in the living factory: metal-accumulating bacteria and their potential for materials science. Proc *Trends in Biotechnol* 19:15-20

[10] Mukherjee P, Ahmad A, Mandal D, Senapati S, Sainkar S R, Khan M I, Ramani R, Parischa R, Ajaykumar P V, Alam M, Sastry M, Kumar R (2001) Bioreduction of $AuCl_4$ Ions by the Fungus, Verticillium sp. and Surface Trapping of the Gold Nanoparticles Formed. *Angew Chem Int Ed* 40: 3585-3588

[11] Ahmad A, Mukherjee P, Senapati S, Mandal D, Khan M I, Kumar R, Sastry M (2003) Extracellular Biosynthesis of Silver Nanoparticles using the Fungus *Fusarium oxysporum. Colloids Surf B Biointerfaces* **28**:313-318

[12] Shiv Shankar S, Ahmad A, Pasricha R, Sastry M (2003) Bioreduction of Chloroaurate ions by Geranium leaves and it's Endophytic Fungus yields Gold Nanoparticles of Different Shapes. *J Mater Chem* 13: 1822-1826

[13] Begum N A, Mondal S, SaswatiBasu, Laskar R A, Mandal D (2009) Biogenic Synthesis of Au and Ag Nanoparticles using Aqueous Solutions of Black Tea Leaf Extracts. *Colloids Surf B Biointerfaces* 71:113-118

[14] Philip D (2009) Biosynthesis of Au, Ag and Au–Ag nanoparticles using edible mushroom extract. *Spectro Chim Acta Part A* 73:374-381

[15] Badri Narayanan K, Sakthivel N (2008) Coriander Leaf Mediated Biosynthesis of Gold Nanoparticles. *Mat Lett* 62:4588-4590

[16] Awwad A M, Salem N M, Abdeen A O (2012) Biosynthesis of Silver Nanoparticles using *Olea europaea*Leaves extract and its Antibacterial Activity. *Nanosci Nanotechnol* 2:164-170

[17] Youyi Sun, Zhi Zheng, Qing Yan, Jiangang Gao, Hongfang Jiu, Qijin

Zhang (2006) Effects of Ag colloidal nanoparticles on luminescent properties of Eu(III) β-diketone. *Mat Lett* **60:** 2756-2758

[18] Kuladeep R, Jyothi L, Prakash P, Mayank Shekhar S, Durga Prasad M, and Narayana Rao D (2013) **Investigation of optical limiting properties of Aluminium nanoparticles prepared by pulsed laser ablation in different carrier media**. *J Appl Phys* **114**:243101

[19] Vigneshwaran N, Kathe A A, Varadarajan P V, Nachane R P, Balasubramanya R H (2007) Silver protein (core shell) nanoparticle production using spent mushroom substrate *Langmuir* **23:**7113–7117

[20] Katrin Quester, Miguel Avalos-Borja, Ernestina Castro-Longoria (2016) **Controllable Biosynthesis of Small Silver Nanoparticles Using Fungal Extract.** *J Biomater Nanobiotechnol* **7:** 118-125

[21] Raheman F, et al (2011) Silver nanoparticles: Novel antimicrobial agent synthesized from an endophytic fungus *Pestalotia* sp. isolated from leaves of *Syzygium cumini* (L). *Nano Biomed Eng* **3:**174-178

[22] Santhoshkumar T, Rahuman A, Rajakumar G, Marimuthu S, Bagavan A, Jayaseelan C, et al (2011) Synthesis of silver nanoparticles using *Nelumbo nucifera* leaf extract and its larvicidal activity against malaria and filariasis vectors. *Parasitol Res* **108:**693–702

[23] Venkatpurwar V, Shiras A, Pokharkar V (2011) Capped gold nanoparticles as a novel carrier for delivery of anticancer drug: in vitro cytotoxicity study. *Int J Pharm* **409:**314–320

[24] Vahabi K, Dorcheh S K (2014) Biosynthesis of silver nano-particles by trichoderma and its medical applications. In: Gupta VK, Schmoll M, Herrera- Estrella A, Upadhyay RS, Druzhinina I, Tuohy MG, editors (2014) *Biotechnology and biology of trichoderma*. Elsevier, London, UK 393–404

[25] Jeyaraj M, Sathishkumar G, Sivanandhan G, MubarakAli D, Rajesh M, Arun R, et al (2013) Biogenic silver nanoparticles for cancer treatment: an experimental report. *Colloids Surf B Biointerfaces* **106:**86–92

[26] Basha K S, Govindaraju K, Manikandan R, Ahn J S, Bae E Y, Singaravelu G (2010) Phytochemical mediated gold nanoparticles and their PTP 1B inhibitory activity. *Colloids Surf B Biointerfaces* **75:** 405–409

[27] Pradeep T, and Anshup (2009) Noble metal nanoparticles for water purification: A critical review. *Thin Solid Films* **517:**6441-6478

[28] Sathyavathi R, Bala Murali Krishna M, and Narayana Rao D (2010) Biosynthesis of Silver Nanoparticles Using *Moringa oleifera* Leaf Extract and Its Application to Optical Limiting. *J Nanosci Nanotechnol* **10:**1–5

[29] Feifei Chen, Junwen Cheng, Shixun Dai, Zhe X, and Qinyuan Zhang, Wei Ji, Ruiqin Tan (2014) Third-order optical nonlinearity at 800 and 1300 nm in bismuthate glasses doped with silver nanoparticles. *Opt Exp* **22**:13438-13447

[30] Nabika H, and Deki S (2003) Enhancing and Quenching Functions of Silver Nanoparticles on the Luminescent properties of Europium Complex in the solution phase. *J Phys Chem B* **107:**35

[31] Mock J J, Barbic M, Smith D R, Schultz D A, and Schultz S (2002) Shape effects in plasmon resonance of individual colloidal silver Nanoparticles. *J Chem Phys* **116:**6755-6759

[32] Qingru Wang, Shangxin Lin, Chengguo Ming, Hongyan Zhao, Jiadong Liu, Chao Zhang, Feng Song,

Edwin Y B Pun (2011) Plasmon-enhanced luminescence of Eu complex by using silver nanocubes for different excitations. *Mat Lett* **65:**905–907

[33] Fang X N, Song H W, Xie L P, Liu Q, Zhang H, Bai X, Dong B, Wang Y, Han W (2009) Origin of luminescence enhancement and quenching of europium complex in solution phase containing Ag nanoparticles. *J Chem Phys* **131**: 054506

[34] Novotny L, Hecht B (2006) Principles of nano-optics. Cambridge University, London

[35] Gagandeep K, Verma R K, Rai D K, Rai S B (2012) Plasmon-enhanced luminescence of Sm complex using silver nanoparticles in Polyvinyl Alcohol. *J Lumin* **132:**1683–1687

[36] Ringler M, Schwemer A, Wunderlich M, Nichtl A, Ku rzinger K, Klar T A, and Feldmann J (2008) Shaping Emission Spectra of Fluorescent Molecules with Single Plasmonic Nanoresonators. *Phys Rev Lett* **100:** 203002

[37] Philip D, Unni C, Aswathy Aromal S, Vidhu V K (2011) MurrayaKoenigii leaf-assisted rapid green synthesis of silver and gold nanoparticles. *Spectrochim Act Part A* **78:**899–904

[38] Mirgorod Yu A, Borodina V G, and Borsch N A (2013) Investigation of Interaction between Silver Ions and Rutin in Water by Physical Methods. *Biophysics* **58:**743–747

[39] Preeti Singh, and Jaspal Singh (2013) MEDICINAL AND THERAPEUTIC UTILITIES OF *RAPHANUS SATIVUS. Int J Pl An and Env Sci* **3:**2231-4490

[40] Philip D (2011) MangiferaIndica leaf-assisted biosynthesis of well-dispersed silver nanoparticles. *Spectrochim Act Part A* **78:**327–331

[41] Elemike Elias Emekaa, Oseghale Charles Ojiefoha, Chuku Aleruchib, Labulo Ayomide Hassana, Owoseni Mojisola Christianab, Mfon Rebeccac, Enock Olugbenga Darea d, Adesuji Elijah Temitope (2014) Evaluation of antibacterial activities of silver nanoparticlesgreen-synthesized using pineapple leaf (*Ananas comosus*). *Micron* **57:**1–5

[42] Agasti N, KaushikAmerican N K (2014) One Pot Synthesis of Crystalline Silver Nanoparticles. *J Nanomater* **2:**4-7

[43] Tam F, Goodrich G P, Johnson B R, and Halas N J (2007) Plasmonic enhancement of molecular fluorescence. *Nano Lett* **7:**496–501

[44] George Z, and Barnes W L (2008) Fluorescence enhancement through modified dye molecule absorption associated with the localized surface Plasmon resonances of metallic dimmers. *New J of Phys* **10:**105002

[45] Wang Q, Song F, Lin S, Liu J, Zhao H, Zhang C, Ming C, and Edwin Y B Pun (2011) **Optical properties of silver nanoprisms and their influences on fluorescence of europium complex c.** *Opt Exp* **19:**6999-7006

[46] Pascal A, Palash B, and Lukas N (2006) Enhancement and Quenching of Single-Molecule Fluorescence. *Phys Rev Lett* **96:**113002

[47] Bala Murali Krishna M, Praveen Kumar V, Venkatramaiah N, Venkatesan R, and Narayana Rao D (2011) Nonlinear Optical Properties of Covalently Linked Graphene-Metal Porphyrin Composite Materials. *Appl Phys Lett* **98:**081106

[48] Shukla V, Singh C P, Mukherjee C, and Bindra K S (2013) Investigation of optical limiting in Cobalt nanoparticles synthesized by laser ablation. *Chem Phys Lett* **555:**149-153

[49] Ramya E, Jyothi L, Narayana Rao D (2018) Influence of Optical Properties of Ag NPs from Raphanussativus Leaf Extract on Lanthanide complexes. Plasmonics 12:1601

[50] Felicia T, Glenn P, Goodrich Bruce R J, and Naomi J H (2007) Plasmonic Enhancement of Molecular Fluorescence. *Nano Lett* **7**:496-501

[51] Fang H, Song H, Xie Liu Q, Zhang H, Bai X, Dong B, Wang Y, and Han W (2009) **Origin of luminescence enhancement and quenching of europium complex in solution phase containing Ag nanoparticles.** *J Chem Phys* **131:**054506

Chapter 4

Polymer Architectures for Optical and Photonic Applications

Ana-Maria Albu and Vlad Marian Târpă

Abstract

The last decade of the last century is marked by a revolution in the synthesis of materials for optical and photonic applications, against the background of the growing need for new high-performance materials to increase the efficiency, reliability and speed of response linked to environmental aspects. The diversity of requirements and the optimization of the responses has led to a major dispute over the structure and composition of these materials: Inorganic or Organic, Natural or Synthetic, Hybrid or Pure, which has stimulated interest in the development of various architectures. Special attention shall be paid to establishing a fundamental relationship to correlate the non-linear optical response and chemical structure, especially for the category of organic materials- particularly polymers- distinguished by structural/compositional versatility and suitable for processing by simple technique which allows serial production. In fact, optical nonlinearity (NLO) is not an exotic phenomenon. Indeed, all materials are optically nonlinear if light is sufficiently intense. The synthesis of functional photonic organic materials is a major challenge of contemporary community of material scientists to imagine new functional materials based of "collective" phenomena by virtue of the "engineered" molecule- molecule interactions and spatial relationships. In this context, this paper aims to highlight the most important features concerning the structural - compositional relationship of polymeric materials used in optoelectronic and photonic applications.

Keywords: polymer, azo-polymer, optical material, azo-chromophor, non linear optics, solvatochromic, photochromic, self anssembly

1. Introduction

Presently, one can remark a continuous increase of the need for new technologies adjusted to the market request. This involves obtaining new high-performance materials and new microelectronic, optical devices, whence appearing these requests. Because we begin will be habitually with these advantages, our expectations demand a higher efficiency, faster and more reliable materials answers, safe environmentally technologies.

Optical nonlinearity (NLO) is not an exotic phenomenon. Indeed, all materials are optically nonlinear if light is sufficiently intense.

The last decade of past century was marked by a revolution in the field of materials synthesis for optical and photonic applications. It is noteworthy major dispute regarding the origin of these materials: inorganic or organic, natural or synthetic, hybrid or pure?

Thus, the fundamental problem in the development of this class of materials still remains the formulation of a theoretical model, which generalizes the particularities of interaction of the stimulating electromagnetic field with the specific micromolecular constituents for each class of compounds.

In this regard, perhaps the greatest challenge is (outside the design synthesis, characterization and understanding of novel molecular and macromolecular assemblies with applicability in non-linear optical fields) the interfacing nature of research, which brings together specialists in chemistry, physics and materials science. Initially, a test stone for chemists, photonics materials require the rational construction of supramolecular assemblies characterized by preordained collective phenomena consequence of intermolecular interactions and their spatial relationships.

It is classic that the high values of the electronic polarization are characteristic for π-conjugated molecules. This justifies the many generalizing predictions that material systems with extended π- conjugation will be characterized by high values of hyperpolarizabilities.

2. Principles of the non-linear optical media

The basic nonlinear optical polarization can be studied at the molecular level. That involves an effort devoted to understanding the relation between the molecular structure and materials properties. Another perspective regards the discussion of the effective arrangement of the nonlinear molecules into macroscopic responses (single crystal, quasi-crystalline thin film, and a polymer matrix, etc.). To describe the nature of nonlinear optical response in materials is necessary to consider the aspects of frequency dependence and symmetry.

A molecular medium, such as an organic crystalline or polymeric solid, is generally non-conducting and non-magnetic and the electrons are regarded as being tightly bound to the nuclei. For such media, the interaction with light can generally be regarded within the framework of a dielectric subjected to an electric field [1–3].

At molecular level, the interaction of molecule with electric field can be described by the polarizability (p):

$$p = \alpha E + \beta E^2 + \gamma E^3 + \ldots \qquad (1)$$

where: α, β, γ are the molecular polarizabilities (1st, 2nd, 3rd order).

The fundamental relationship describing the induced charge in molecular dipole moment (polarization) upon interaction with an oscillating external field (light) can be expressed in a power series:

$$p_i = \sum_{j} \alpha_{ij} E_j + \sum_{j \le k} \beta_{ijk} E_j E_k + \sum_{j \le k \le l} \gamma_{ijkl} E_j E_k E_l + \ldots; \qquad (2)$$

here p_i is the electronic polarization induced along the ith molecular axis, E_j is the jth component of the applied electric field, α is the linear polarizability, β is the quadratic hyperpolarizabilitiy and γ is the cubic hyperpolarizability. Starting with the second term of equation are described the nonlinear response to the radiation field and account for terminology "nonlinear optics". Importantly, the even order tensor, β is responsible for doubling frequency of incident light (second harmonic generation, SHG) and γ describe the third order response of material at molecular level (third harmonic generation, THG).

At macroscopic level, nonlinear optics is concerned with the electric polarization P (dipole moment per unit volume) induced in a material by an electric field E.

Typically, P is proportional to E and oscillates at the same frequency. *So, the response is linear!* As the magnitude of the field increases, P deviates from proportionality and nonlinear phenomena appear:

$$P_i/\varepsilon_0 = \sum_j \chi_{ij}^{(1)}{}_{ij} E_j + \sum_{j \le k} \chi_{ijk}^{(2)} \beta_{ijk} E_j E_k + \sum_{j \le k \le l} \chi_{ij}^{(3)} E_j E_k E_l + \dots \quad (3)$$

$\chi^{(n)}$ = electric susceptibilities of order n; linear for n = 1, quadratic: n = 2, cubic: n = 3.
Considering the dual character of light, electric and magnetic:

$$E = E(\omega) + E(\Omega) = E_0 \cos(\omega t) + E(\Omega), \quad (4)$$

where $E(\omega)$ is the optical component of field and $E(\Omega)$ is characteristic for external applied field, can be written the mathematical expression of polarization as function of frequency:

$$P(\omega) = \left(\chi^{(1)} + 2\chi^{(2)}E(\Omega) + 3\chi^{(3)}E(\Omega)^2 - {}^3\!/_4\chi^{(3)}E_0^2\right) E_0 \cos(\omega t), \quad (5)$$

where: the second term describe the Pockels or linear electrooptic effect (1th order) the third is for the Kerr of quadratic electrooptic effect (2th order), and the last is specifically for optical Kerr effect.

$$P(2\omega) = \left({}^1\!/_2\chi^{(2)} + {}^3\!/_2\chi^{(3)}E(\Omega)\right) E_0^2 \cos(2\omega t), \quad (6)$$

where the first terms specifically for second harmonic generation (SHG) and the second describe the electric-field induced SHG (EFISH),

$$P(3\omega) = \left({}^3\!/_2\chi^{(3)}E(\Omega)\right){}^1\!/_4\chi^{(3)}E_0^3 \cos(3\omega t)E = E_0 \cos(\omega t) + E(\Omega), \quad (7)$$

which is specifically equation for third harmonic generation (THG).

Starting from this fundamental description of interaction light-material, without forgetting about the multicomponent structure of materials (different polar species), can be formulate two fundamental aspects for the researchers interested by the synthesis of such materials:

- The design of this materials suppose an biunivocal interdependence of macroscopic and microscopic optimization;
- If a microscopic level is very important to enlarge β and γ values, at macroscopic level very important is the effective arrangement of the nonlinear molecules into macroscopical response. So macroscopic susceptibility describe the nature of nonlinear optical response in materials; considering aspects of frequency dependence and symmetry. Generally:

$$\chi^{(2)} = Nf\beta, \text{respectively}\, \chi^{(3)} = Nf\gamma, \quad (8)$$

where N is the number of molecular dipols (polar or polarizable molecules), f is the local field factor.

Synthetic materials of interest in ONL applications are not isotropic. For example:

- Organic crystals consist of molecules with specific orientations of structuring the constituent motif. Thus, the molecules themselves often have electronic structures with a high degree of anisotropy and consequently, the response of the material to the applied electromagnetic field will not always respect the direction of incidence, as is usual in classical optics.

- Similarly, in terms of orientation, polymeric chains comply with the principle of minimum energy. If these materials are generally characterized by the random orientation of chains - isotropic structure - depending on the composition, the chains constituting them may be oriented preferentially, so that their energy distribution will require their response to external stimuli.

For this reason, susceptibilities are tensor quantities related to the polarization response in one of the three directions of the field. The material characteristic that reflects the degree of anisotropy is the dielectric constant ε_{ij}, defined by the electric field and the dielectric displacement, being a second-degree tensor. The birefringence phenomenon is one of the consequences of this dielectric anisotropy. Visualizing a uniaxial crystal with nx = ny + nz, the refractive index of the materials, n, is given as

$$n \approx \sqrt{\varepsilon} \tag{9}$$

and with a light wave propagating from left to right along the x-axis with its electric field polarized in the z-direction, the propagation constant k_z will be determined by n_z from:

$$k_z = \frac{\omega}{c} n_2 \tag{10}$$

Thus, while all material structures from the point of view of the optical properties, are characterized by the presence of the both hyperpolarizabilities (β, γ), do not display simultaneous second and third order nonlinear response. That is due to the peculiar structural requirement for the two NLO properties. The high concentration of chromophores with large product of the dipole moment, μ, and the first hyperpolarizability, β, stimulate aggregation tendency, diminishing their contribution to the electrooptic effect of materials. The last decade of the 1990s has established that high electro-optical coefficients for guest-host material systems, with polymeric matrix, are essentially due to electrostatic chromophore-chromophore interactions. Thus, when doping the polymer matrices with chromophe species characterized by high electrostatic interactions, there is a decrease of the electro-optical coefficients in the value with an increase of the chromophore concentration. As well, the poling efficiency is limited because a disrouption in alignment of chromophes appears when the polarization field is stoped, owing to the electrostatic interactions and thermal randomization. This behavior is the result of a major relaxation phenomena of chromophors' in the first moments after stopping the polarization field [4, 5].

The improvement in the value of electro-optical coefficients requires intervention in the composition and structure of the chromophore, without perturbation of of the π-electron structure to avoid changing the molecular hyperpolarizability β. In such cases, the syntheses that favor the perpendicular orientation of constituent segments with distinct polarity (i.e. acceptor, donor) to the chromophore conjugation system, are preferred. Such sterical hidrance limits the distance at which chromophoric molecules can approach each other, thus reducing undesirable dipole–dipole interactions; finally becoming competitive with lithium niobate [6].

2.1 Origin and maximization of molecular hyperpolarizabillities $\chi^{(2)}$, $\chi^{(3)}$

Early on, the correlation of the β size value with the permanent dipole moment difference, between the ground and excited electronic state, underlines the dependence of the SHG signal magnitude on the material structure. Although considerable efforts have been made to create new "push-pull" conjugate sequences, that generate a large charge in the excited state, the predicted response values have not always been obtained.

The correlation of the material response with the extension of double/single bonds alternation in the polyene structures served as a model in development of an elegant model for understanding the relationship of the chemical structure-nonlinear optical response. The model is still a useful guide to design the materials with predicted SHG properties, highlighting, as a key element in generating the NLO response, the size of the electronic conjugation at the level of the donor-acceptor substituted chains [7–15].

Centrosymmetric media are materials with a high density of polarizable electrons, which stimulates the apparition of strong third order NLO phenomena.

Increasing the magnitude of NLO phenomena in polarizable polymers involves either maximizing the intensity of the applied polarization field or designing molecules with intrinsic steric hindrance and an adequate maximization of electronic conjugation at level of the substituted donor-acceptor sequence [16–18].

As extended conjugation is a key element for third-order NLO materials, several constructive restrictions need to be observed [19]:

- high molecular symmetry of the material sequences that generate the excited and de-excited states, respectively;
- involving the high energy excited state in the material response;
- a large molecular dipole moment variation for transition ground state to excited state.

Theoretical and experimental paper of the field highlights that the best candidates for these applications must be characterized by a high level of extended conjugation to the chain level as well as by strong monophotonic absorptions. Thus, structures similar to polyene compounds are ideal, such as: poly (di)acetylenes, polyarylenes or poly (vinylarylenes) as good as the aromatic structures of chinoids (push-pull). [10, 11, 20–23].

The length of the conjugate sequences as well as the molecular geometry and the electrical charge of the ground state are the key factors for the type of application of the material.

3. Structural features of NLO polymer materials

In the previously outlined context, the researchers' attention was focused on studies to generalize the NLO response-chemical structure correlation, while spending efforts on applied research aimed at creating new photonics and photoelectric devices. The variety of physicomechanical properties as well as the compositional versatility of conjugate organic materials placed them at the center of these studies. The π conjugate organic materials with large electronic polarizabilities and, since the nonlinear response is a weighted summation of higher molecular polarizabilities, it becomes reasonable for conjugate molecules to possess substantial hyperpolarizabilities.

3.1 NLO chromophors

The researches activities in the materials fields as well as application development mainly based on multiphoton absorption (MPA) process continue with consistency to demonstrate the successful proof-of-concept in diverse areas, such as microfabrication, bioimaging, photodynamic therapy, frequency upconverted lasing, etc. The mid-1990 is marked by tempestuously development in the design and synthesis of the new dyes, which are more efficient than those commercially available. Consequently, in the coming decades it became evident that these together with theirs processability, photostability, and durability, that depends on the application, are the basic elements for the specificity of physical motif. The great continuous challenge for technological exploration remain how to establish and fine-tune structure-properties relationships for these large numbers of organic, organic–inorganic materials assemblies, taking into account to the varied molecular structural factors preserving the reproducible NLO properties. It was known that molecular hyperpolarizability increases with the π-electron conjugation length of the chromophore. Although it would be expected that the increase in chemical reactivity and optical absorption would be directly proportional to the length of the conjugation at the chromophore level, it has experimentally proved that it is not the best approach to optimize nonlinearity. Theoretical models developed for molecules prone to charge transfer show that the strength and energy balance of donor-acceptor sequences, correlated with the length of the electronic bridge, directly affect the value of molecular hyperpolarization [24–26]. Thus, by adjusting the strength of the donor and acceptor sequences, the moment of the fundamental state, respectively excited, is directly determined. These peculiarities claim that the increase of the value β is the consequence of the mesomeric interactions of the strong donor-acceptor groups. In conclusion, in the case of organic compounds, the structure-properties relationship is closely related to the strength of the D-A bond, as well as to the widening of the conjugation in the structural unit:

$$\beta \approx \left(\mu_{ee} - \mu_{gg}\right) \mu_{ge}^2 / E_{ge}^2,$$

where: g = ground sate index; e = excited state index; μ_{gg} = ground state molecular dipole; μ_{ee} = excite state molecular dipole; μ_{ge} = transition state molecular dipole; E_{ge} = transition energy.

Pursuant to this equation, hyperpolarizability is directly determined by the change of the dipole moment and the oscillator power (expressed by the square of the matrix of the dipolar elements) and is inversely proportional to the square of the transition energy. Each of these parameters has a maximum value closely related to the chemical structure of the molecule.The specific function (electronic exchange) which determine the NLO character, as well as the strength D-A, shall decided by the variation of all mentioned factors.This dependence can be found in the dominant β value, approximate by Hückel series for molecular orbitals calculation.

Thus the coulombic energetical difference between donor (α_D) and acceptor (α_A) can be explained as energetical HOMO-LUMO dissimilarity. The parameters $\mu^2{}_{ge}$ and $E^{-2}{}_{ge}$ show maxima value for the null molecular asymmetry, so: $\mu_{ee} - \mu_{gg} \approx \alpha_D - \alpha_A$ (**Figure 1**) [8]. Consequently, at increasing of molecular asymmetry $\alpha_D - \alpha_A > 0$, μ_{ee}-μ_{gg} increase with the difference α_D-α_A. Increasing the α_D-α_A, the mixing of bridge orbital's D-A decrease in the same time with transition HOMO-LUMO. At the same time, there is a decrease in the charge transfer character, specific to stilbenic molecules substituted with D -A groups.

The D-A molecular structures are limited on the one hand by the coulombic factor (which may have negative, positive or zero values depending on the level of charge separation in the molecule) and on the other hand by the resonance energy, the consequence of the aromatic electrons involved in the conjugate DA.

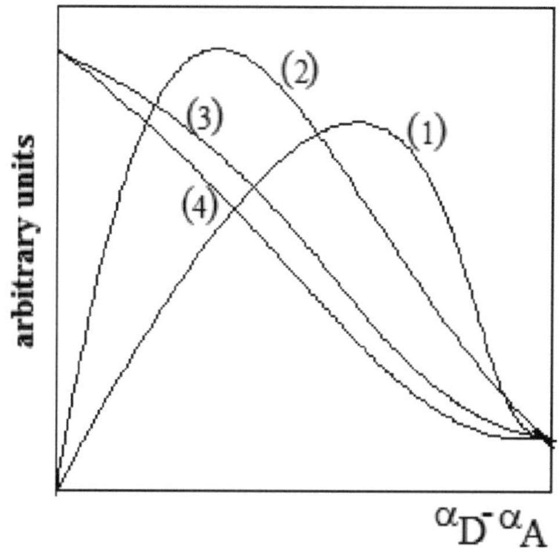

Figure 1.
The dependence of μ_{ee}-μ_{gg} *(1),* β *(2),* μ_{ge}^2 *(3), and* $1/E^2_{ge}$ *(4) as function to* $\alpha_D - \alpha_A$.

In addition, the dipolar transition moment show a sinusoidal variation for β with D-A strength while maintaining a consistent bridge of the conjugate bridge (**Figure 1**). Chromophores that were initially used in the NLO study were of limited success as they failed to reach the required absorption maxima in the near IR wavelength range (**Table 1**). There is a strong correlation between intramolecular charge transfer processes and NLO response, if discussed in terms of electronic structures and physical processes. Therefore, the design of an ideal chromophore structure with high NLO activity is based on some fundamental aspects at the molecular level. In this idea the intermolecular charge transfer, as a driving force, is necessary but not enough, so it is clear that there must be a high electronic density in the material. On the other hand, another fundamental problem in the operation of NLO materials is the length of the conjugate (i.e., the path length), whit determine the hight and permanent charge separation at molecular level [18, 29–35].

At the outset of reviling the particularities of molecular structures with NLO properties, two general types of organic molecules predisposed for these phenomena are defined (**Figure 2**) [36]:

a. The Type I Cromophers have a symmetrical structure and, in general, are made up of an aromatic bridge, rich in electrons, flanked on both sides either by deficit heterocyclic groups or by electronic surplus;

b. Type II chromophores are asymmetric molecules formed by an aromatic/olefinic bridge, rich in electrons, flanked by an electro-donating group, respectively, electron acceptors.

Improved solubility required the attachment of pendant alkyl chains to the aromatic sequence. At the same time, the attachment of electron-donor/acceptor groups models the intrinsic electronic density of high molecular weight polymers. The criteria for achieving symmetrical structures (type I (a) and I (b)) are based on the correlation between the large loads of the quadrupole moment during photoexcitation and the cross section. Consequently, molecules characterized by the alternation of vinyl and 1,4-arylene groups as well as the connectors between two identical end groups (i.e., electron- acceptor or donor) are preferred to realise of many NLO chromophores. Such structures are described generically as “push-push”

NLO chromophore	$10^{-30}\beta\mu$ (cm^5/esu)	NLO chromophore	$10^{-48}\beta\mu^{*}$ (cm^5/esu)
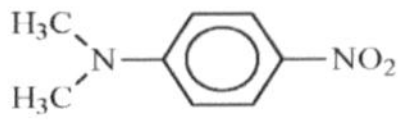	138	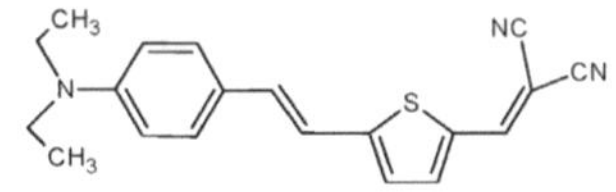	1 300
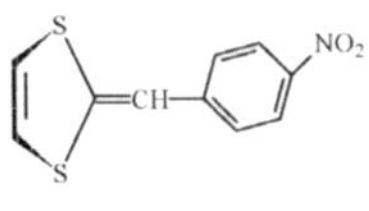	358	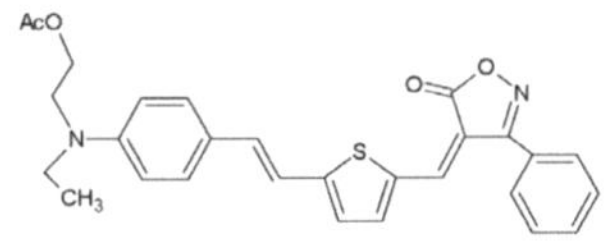	2 000
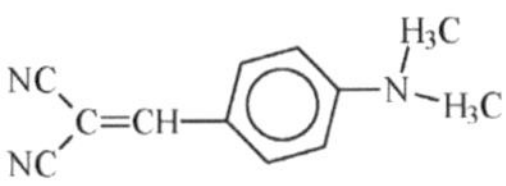	271	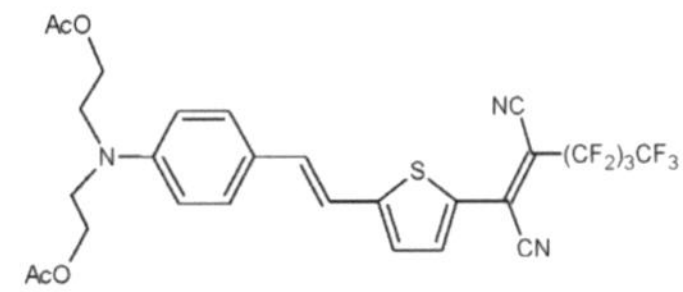	3 300
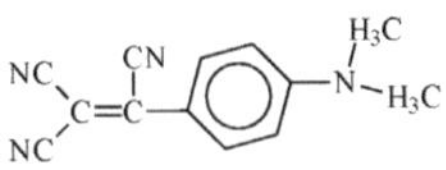	846	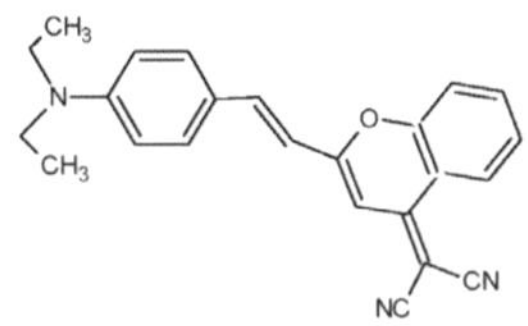	1 720
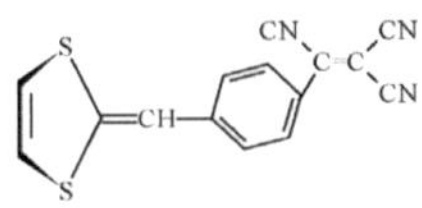	1 200	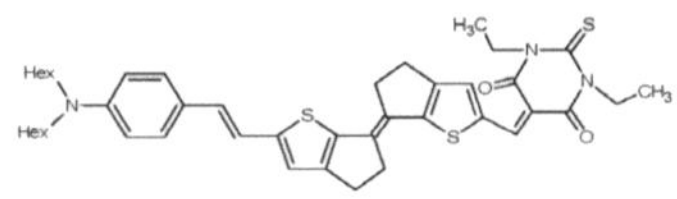	10 400
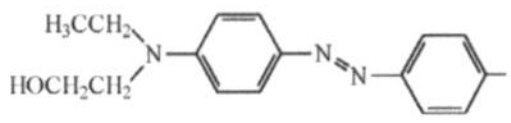	1 090	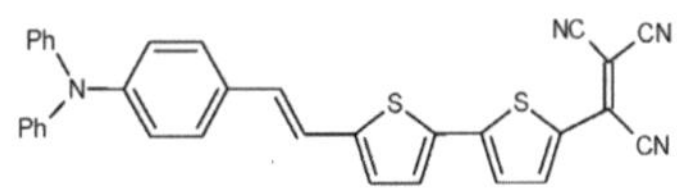	10 600
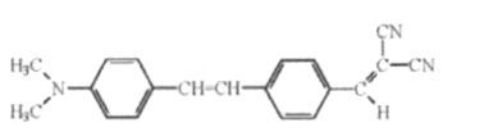	2 650	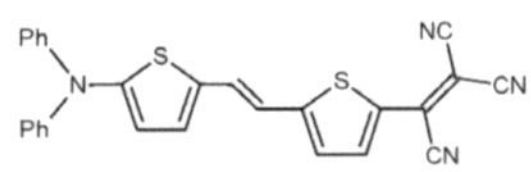	10 200
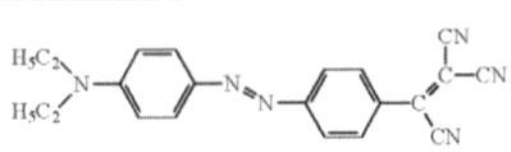	4 110	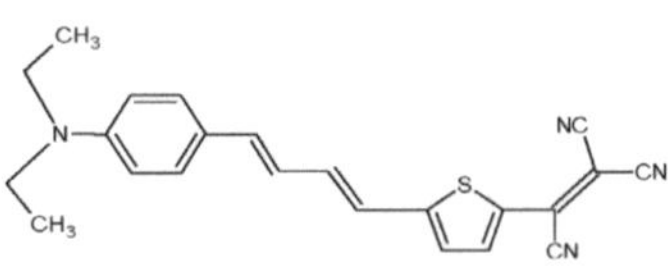	9 800
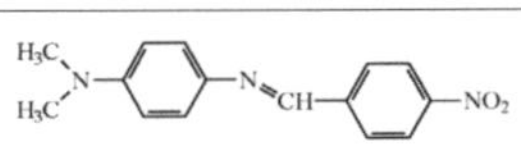	500	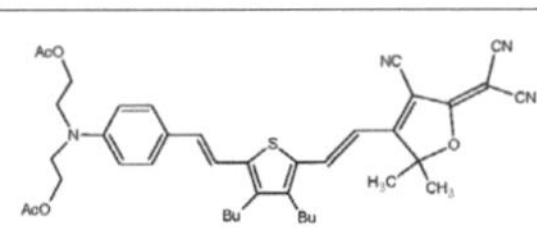	18 000
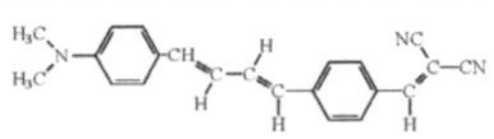	1 320	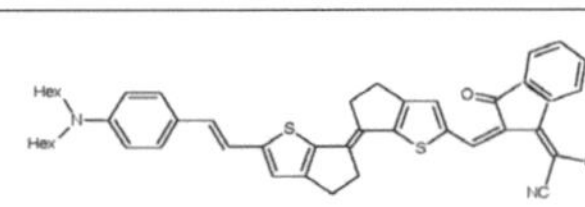	19 400

*At 1907 nm.

Table 1.
Most investigated NLO chromophores and their βμ-values [27, 28].

(I (a)) or "pull-pull" (Type I (b)), depending on the direction of intramolecular load transfer: from the end to the centre of the molecule, or vice versa. This model has subsequently gained growing popularity, proving very useful in the design of one-dimensional molecules [29]. The molecular coplanarity is another key parameter in enhancing the intramolecular charge transfer efficiency [37]. At the same

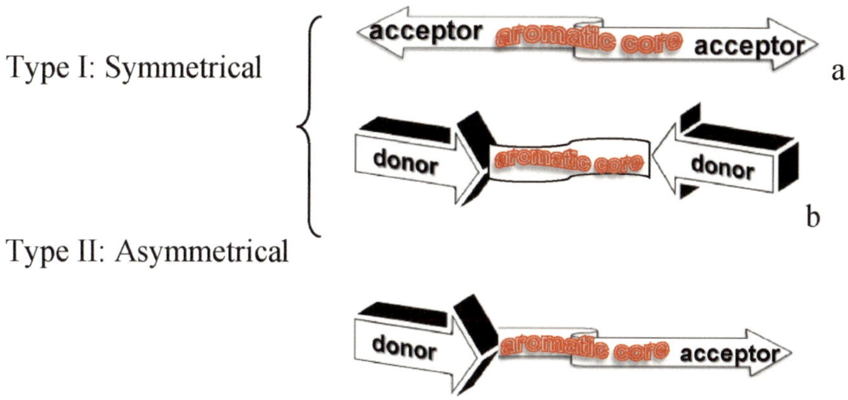

Figure 2.
Classical molecular structures for NLO chromophors: Ar aromatic block.

time, for organic molecules, the strength of dipole in the ground- state of nonsymmetric molecules as well as the multipolar transition-dipole strength in the centrosimetric molecules are also energy factors that determine the NLO response magnitude [29, 38, 39]. Both experimental and theoretical, it has been demonstrated that the multiplication of the number of conjugation path or the connection of more than one linear path to generate two- or three-dimensional configurations, leads, in the material, to a major increase of the NLO responses value [31, 32, 38, 40].

Remarkable is the fact that when assessing the properties of NLOs, the energy changes caused by the interaction with the environment should not be neglected [41]. Thus, the polarity of the solvent, expressed by its dielectric constant, is an important structuring parameter determining on the one hand solvation by hydrogen bonds or by promoting colloidal-microaggregate dissolution. Thus, the efficiency of the cross section is more sensitive to the polarity of the local environment than to the intrinsic NLO aspects (properties) of the material [42].

3.2 Functionalized polymer materials systems

The NLO polymeric films are usually composed of long chains with chromophore sequences. Thus, chromophores, which are seen as net dipoles, are the origin of optical non-linearity in polymers. The interaction between the 'non-polar' molecules from bulk material and the chromophore, the molecular orientation (either random or of a certain symmetry) can introduce an overall orientation of the molecular assembly. This orientation will tend to cancel the individual molecular contributions. It is therefore important to ensure that all molecules are aligned correctly. The first information on this issue can be obtained by analyzing the chromophe sequence.

The starting point in developing technical strategies for the real NLO polymers synthesis is both the design of the assembly at the molecular level and the design and characterization of chromophe molecules with high molecular hyperpolarizabilities (β, γ), able to be, easily and efficiently, incorporated in a polymeric matrix.

The generous literature information, both experimental and theoretical, on the use of organic materials (especially polymers) in NLO applications is based on the composition-properties relationship. A particularly important aspect relates, in addition to practical assembly techniques, to their composition, synthesis and characterization. Therefore, for polymerist researchers, it becomes defining the aspects of the synthesis and reactivity of the monomers bearing NLO sequences as well as the establishment of techniques for incorporating these NLO sequences into the global polymeric architecture.

A particular aspect in the synthesis of NLO polymers, beyond optimization of the individual response of the constituent chromophore, is how the chromophore density can be maximized while preserving micro- and supra-structuring properties, respectively. The experimental studies, developed in this context, outlined two alternative strategies (**Figure 3**):

1. A first alternative is offered by the chemical modification of preformed macromolecular chains with chromophore sequences suitable for the desired application. This path offers the advantage of using chains with dimensions useful to the target application. The method is limited to relatively low degrees of transformation, does not provide a rigorous control of the distribution of the attached chromophore and last but not least, involves a series of purifications subsequent to the synthesis, in order to remove the unreacted chromophore.

2. However, a macromolecular structure with a predefined architecture can be easily obtained by (co) polymerization of monomers carrying an electrooptically active sequence. The major advantage of this method is the rigorous control of the chromophore fraction in the macromolecular structure both as a weight and as a sequential distribution. The only restriction imposed concerns the intrinsic structure and purity of the monomers which should not impede the (co) polymerization process.

Based on theoretical prediction and quantum the most promising macromolecular materials were, initially, the electron -conductive polymers such polyacetylenes, polythiophene or poly(p-vinyl phenylene)'s. Few research groups underlined that the lengths of conjugate sequences is not a necessary preconditions for increasing the intern polarity value (implicit $\chi^{(2)}$, $\chi^{(3)}$). Therefore the oligomers segments can be understood as inactive spacer with broaden the NLO activity,

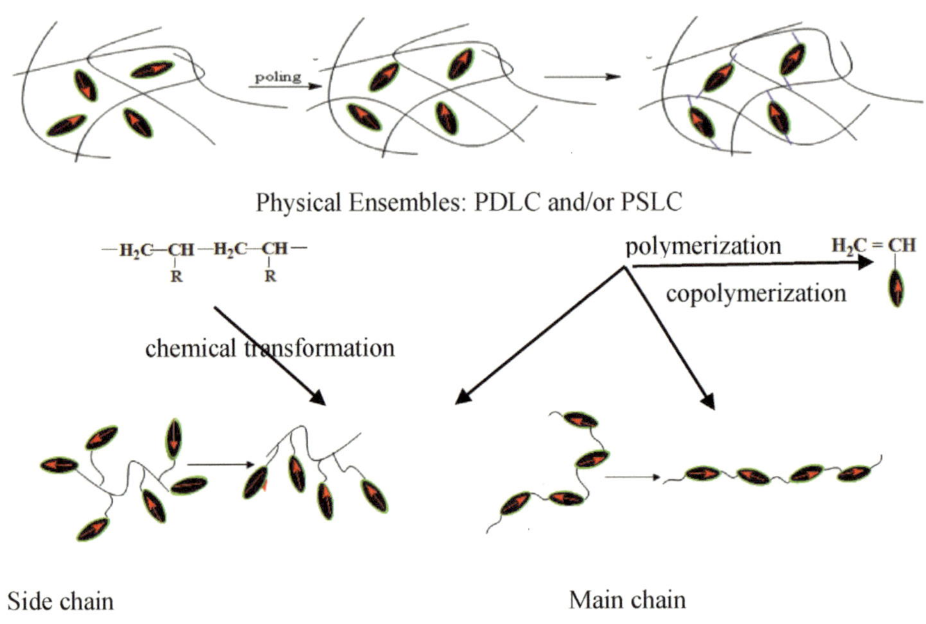

Figure 3.
Design Motifs of NLO polymers materials.

simultaneous facilitate the processability of materials [43–45]. Moreover, the susceptibility tensor calculus, with theory the disturbing torque- Hertree-Fock, for polyene series $C_{22}H_{24}$, underline the dependence $\gamma_{xxxx} \approx L^4$; (L chain length) [45]. Afterwards were demonstrate that $\gamma_{xxxx} \approx L^{4,6 \pm 0,22}$ [46].

For poly (p-vinyl phenylene), the predictions of the value rapport γ/N, pointing out a decreasing with increasing N(dipols number). In this context, it is outlined the idea that the best polymeric structures are those consisting of conjugated oligomeric segments, connected with spacers with or without NLO activity (**Figure 4**) [22, 47].

Through processing limitations (low solubility in the usual solvents) the researcher works were directed towards finding a new modality for including these sequences in polymer structure. So, the ladder hyperconjugate moieties thiophene oligomers and/or polyamide/imide were attached to the polyenic skeletons. Such materials were characterized by high values of hyperpolarizabillity, $10^{-11} \div 10^{-12}$ esu/cm.

The copolyamides structures with polyvinylthiophene, obtained by interfacial polymerization ($M_n = 10^5$), were characterized by thermal resistance until 400°C. The electro-optical analysis of these films denotes the existence of conjugate phase which generated $\chi^{(3)}/\alpha \sim 10^{-3}$ esu/cm. This technique offers the possibility to tune the NLO activity by the number of the electrooptic segments of different species [48, 49].

The similarity of poly(urea)s properties with NLO "piezoelectrical polymers" (poly(vinyl fluorine), poly(vinyl triflourine), copolymers of the triflourine ethylene and tetra- fluorine ethylene, nylon, vinyl cyanide entitle their NLO activity. The strong interchains bond (hydrogen bonds) rule the dipoles alignment in the electric field, [50].

The d_{33} coefficients evolution highlight a very slow relaxation, emphasize the lead of hydrogen bond for maintaining the NLO properties of the material. Starting with these conclusions great efforts were dedicated to investigation of the polymer structures showed in the **Figure 5** [51]. The NLO properties are due to the junction of the donating electrons (2) with the π electrons system. The flexible chain (σ bonds (3) (**Figure 5**) play an important role regarding the solubility, transparence and fusibility of the polymers.

The constitutive- compositional similitude with polyureeas and the peculiar features: (i) the easy chromophores alignment for each synthesis step, the relative low viscosity; (ii) the high rate of polymerization (few seconds to several minutes) –reaction izocianate/alcohol; stabilization of NLO response by the hydrogen bonds, characteristics for polyurethanes; justifying the investigations of NLO polyurethane comportment [52–54]. The study of stability discloses in this case the importance of simultaneous development of polymerization and poling. So, by this procedure were diminished the phenomena of loss of property to reorientation (keeping 40–60% of the initial value) [54].

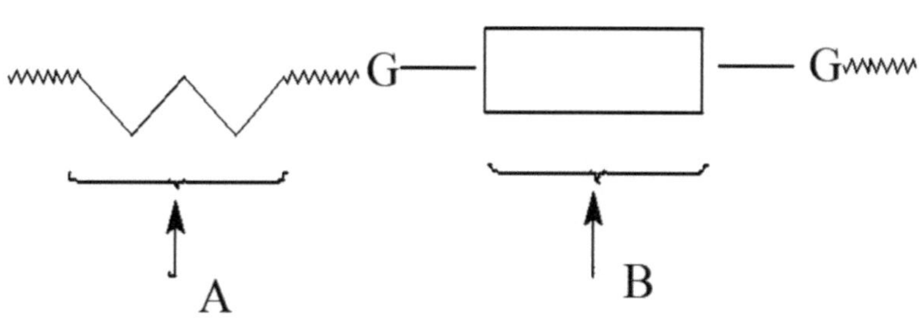

Figure 4.
The structural scheme proposed for NLO copolymers: A = segments with low NLO activity; B = oligomers segments with high NLO activity; G = functional groups [22].

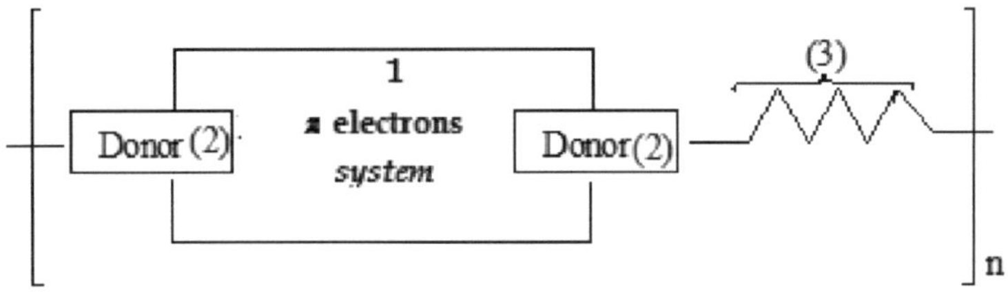

Figure 5.
Macromolecular materials structure used for waveguide fabrication.

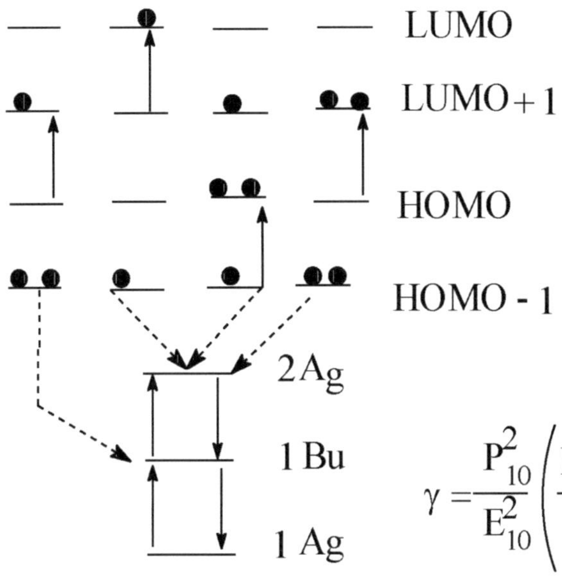

$$\gamma = \frac{P_{10}^2}{E_{10}^2}\left(\frac{P_{21}^2}{E_{20}^2} - \frac{P_{10}^2}{E_{10}^2}\right)$$

Figure 6.
Representation of the two excited states which contribute to the γ value: E: energy of band gapes; P: transition dipole moment; for conversion the three electronic states are denoted as: 0 = G(1Ag); 1 = 1Bu; 2 = 2Ag [51].

The contribution of π-electron sequence, their length, is not crucial for increasing the hyperporarizability value (specially $\chi^{(3)}$). In this case significant is the sequence planarity. On the other hand, the symmetrical substitution, broadening results the enlargement of the transition bands fromHOMO-1 to HOMO and similarly from LUMO to LUMO-1. These transitions promote the third order susceptibility, theoretical prevue by relation showed in the **Figure 6**.

A special attention was granted to the epoxy macromolecular structures. These investigations are supported by the potential stabilization of the chromophore, by its covalent binding to the carrier matrix [55–57]. Another very important aspect, especially if the polymer precursor has a high reactivity, was giving rise to the crosslinked structures. The major advantage, in these cases, is using an amine chromophore monomer. So, the dipoles chromophores alignment is accomplished by crosslinking control simultaneous with polymerization process. A wide range are getting the epoxydic compounds, which is a gain for the electrooptics coefficients (10–50 pm/V) [58–63].

Also of interest for the NLO has proved to be polysiloxane structures (**Figure** 7). This orientation is justified by their versatility, ability to form the films, very useful for monolayers deposition (Lagmuir-Blodgett) and the possibility of obtaining the self- stabilized dipolar structures [64–66]. The assessment of the SHG value, for all

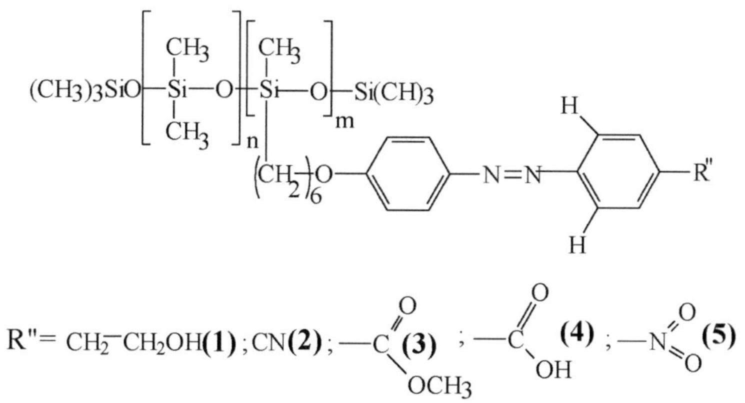

Figure 7.
General structure of polyorganosiloxanes for NLO applications: R", m = 6 ± 2; n = 8 ± 2 (R and R' are hydrogen's atoms) [65].

monolayers, highlights a significant increase while increasing the substituant polarity. This suggests that the package of each polymer chain in the monolayer is orderly, and the interchain bonds are essential in achieving the non-centrosymmetric structure.

Perhaps the most investigated NLO material class is azobenzene derivatives, known and studied for a very long time, their initial use being dye for a variety of substrates. Azobenzenes Photoisomerization is a well-known phenomenon, dedicated to reference materials [67, 68].

The key to exploiting the unique behavior of the azobenzene sequence is defined by how it can be included in the functional materials. For the generation of new properties, the most effective method of incorporation of azobenzene is the covalent attachment to polymers. The materials thus obtained will show both properties specific to polymers: inherent stability, rigidity and processability as well as conformational change under the action of electromagnetic radiation, specific to azo segments - photoisomerization processes. The covalent incorporation of the azo sequence in the polymer, through its photoisomerization process can generate a wide range of phenomena, even unexpected ones [69].

About 25 years ago, at dispersing of azobenzene groups in the polymer matrix, the unique phenomenon recorded was associated with their photoisomerization [70].

Under the action of polarized laser radiation, in polymer films with azobenzene sequence, their orientation is perpendicular to the polarization direction. Dichroic phenomena are thus generated.

This type of phenomenon is reported in the years '80 in liquid crystal polymers with photoactive azobenzene mesogenic sequences [71, 72]. Simultaneously, in Japan, the "command surfaces"concept [73] is opened, in which the azobenzene group acts as 'commander' and the Langmuir — Blodgett liquid films play as 'soldiers', which are aligned in trans/cis conformations by azoderivatives —the switching initiated by illumination [69]. After 1991, the applicative exploitation of these phenomena expanded worldwide, highlighting interesting and unexpected phenomena, which diversify use in reversible optical memory applications as well as photonic applications. Over the last 40 years, an exponential development of the domain has been observed and there is still no analysis covering all these new implications of azobenzene photoisomerization in polymeric structures. Remarkable and still of real interest remain the reviews which treat, generally or in a specific way, phenomena generated by the presence of azoderived sequences

[74–78]. However, the best available analysis covering both photoinduced birefringence and surface relief grating dates back to 2000 [79].

The special interest in azobenzene polymers is the result of their particular properties, which are useful in the processing and storage of optical information [69, 80–82]. The sin-anti conformational conversion process at the bond N = N, presumes some distinct transformations, highlighted in **Figure 8a)**: (i) decoupling of the π electrons from the double bond, (ii) rotation around the single bond N-N followed by (iii) restoring the double bond, N=N, in a new configuration. Assisted by the adequate wavelength radiation, this phenomenon is governed by electronic transitions π-π*, respectively n-π [23].

The molecular photoorientation of chromophe sequences is governed by anisotropic absorption, so that under the action of incident radiation, the dipole moment of chromophe sequence aligns along the molecular axis (**Figure 8b**). The linear polarized light is capable to induce a centro-symmetrical orientation in which the azoic sequence is directed perpendicular to the light plane. These oriental oscillations of the dipole moment's are transferred to the macro-scopical level, stimulating the emergence of optical dicroism and birefringence.

Therefore, the energy metastability of the sin-isomer plays a decisive role in both the theoretical interpretation and the practical application of the photoinduced phenomena in azoic polymers. For example, applications that rely on the modulation of the molecular alignment of CL will use materials with high stability of the sin-isomer (holography, optical data storage) while operational devices based on optical birefringence and dicroism phenomena will only use structures with rapid reversion of the sin-isomer.

The Red Dispers-based materials are the most representative in the class of polymers with the azobenzene side-chain NLO sequence [83–87]. Although noted by good temporal stability of the orientation (preserving about 80–90% of the maximum property value at 1000 hours after orientation) and, in addition, they admit a wide range of refractive index values from UV exposure, a still ongoing problem remains the stabilization of their orientation through photo crosslinking [88].

As the free radical mechanism is easily accessible, an informational richness covers the synthesis, characterization and use of (meta)acrylic and styrenic (co) polymers, with side-chain azo sequence (**Figure 9**) [74, 88–98].

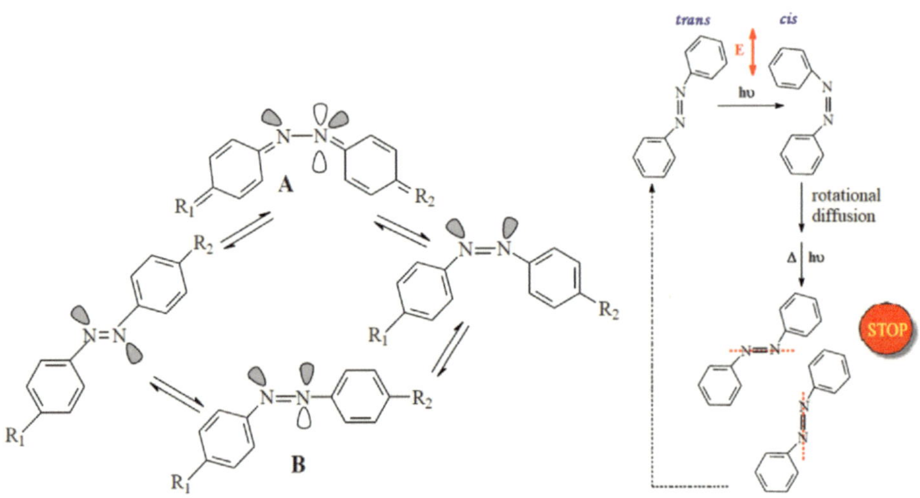

Figure 8.
Conformational changes in azobenzene: (a) A: Rotation of substitutes around–N = N- bond; B: Inversion of substitutes; (b) the azobenzene - light interaction.

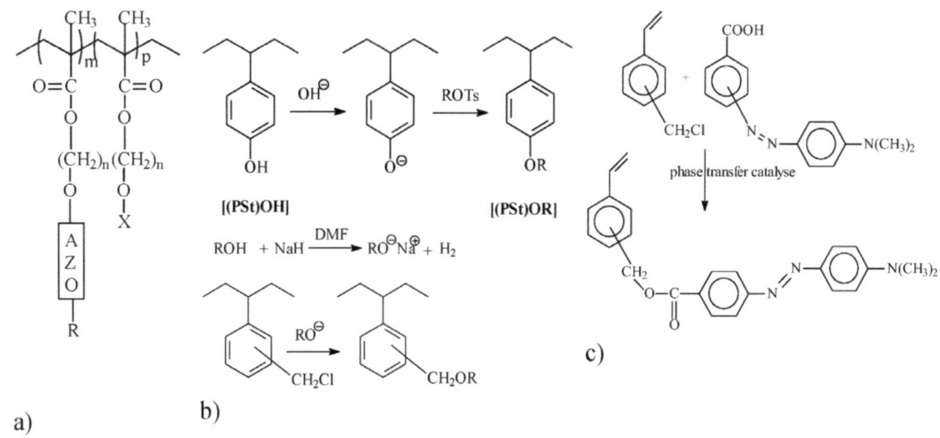

Figure 9.
Structures for methacrylate and styrene materials for NLO applications: a) R = CN; NO_2; $N(CH_3)_2$; OCH_3, CH_3; X = alkyl; Chemical modifications of polystyrene and/or chloromethylestyrene; c) synthesis of NLO styrene monomer structure.

The versatility and easily processing of methacrylic polymers recommend these for NLO applications. The new monomers structures are investigated for chemical reactivity in the (co)polymerization [99–101] as well their electrooptic characterization [102–105]. These research points the inclusion of these cromogene sequences in copolymer structures without major difficulties. A peculiarity of these monomers is their predisposition to an hight chain transfer, which involves the design of syntheses in compositional-operational conditions so as not to drastically reduce the degree of polymerization. In addition, the presence of coloured monomers leads an increase in the rate of polymerization, the phenomenon of self-acceleration becoming obvious. The sequential distribution confirms the predictions obtained from the capitalization of the reactivity ratios and at the same time, the characterization of the material proves the preservation of the optical characteristics of the chromophore.

The styrene materials are preferred relative to methacrylic materials. They are so much easier obtained by chemical modifications (**Figure 9b**): usually is used one of the vinyl benzyle chloride isomers. The drawback of these polymers is the relative rigidity of the chromogen sequence due to the aromatic spacer. The major advantage for these SC structures is the decupling of the chromogene, which facilitates the molecular dipole orientation; but the poling-induced order in organic/polymeric materials is thermodynamic unstable. Thus, an important objective for the NLO polymer synthesis is the structural induction of slow relaxation in the acquired polarization orientation.

A promising alternative is azobenzene SC liquid crystalline polyester architecture [106, 107]. These, by their own nature, extend flexibility at the base-catene level by interactions with both acid and glycolic sequences that can serve as the link sequence of the mesogene. Thus, by modular construction, can be adjusted the length of the flexible methylene spacer in the lateral chain, the substitute on the azobenzoic fragment, the length of the methylene sequence in the main chain (all aliphatic) and the molecular weight of the polyester. Moreover, each parameter obtained by the versatility of the esthetically bond significantly influences the optical storage behaviour of the materials [108]. All materials (with a wide range of substitutes: cyano, nitro, methoxy, hydrogen, methyl, n-butyl, phenyl, fluorine, trifluoromethyl, chlorine, brom) have a diffraction efficiency of over 50%, giving these materials great diffraction properties. The stored information on azo copolymer media (75–100% azo- dye content) can be partially erased up to 80°C [109].

The copolymers of maleamic monomers with styrene and vinyl benzyl chloride and theirs preliminary characterizations of these materials proves that these monomers can be used for build NLO materials [110–112].

A critical issue of materials used in applications based on NLO properties is their thermal stability. This shortcoming can be addressed by promoting (after polarization) cross-linked structures, in particular by photochemical processes. Preferred for this direction are aromatic polyimides (PIm), attached to the polymer frame. The chromophore can be 'grafted' onto PIm sequence through transformation reactions. The method is advantageous because it allows the attachment of a different range of chromophores to polymer supports with a wide compositional variety, promoting the production of materials with high Tg (around 220°C), excellent solubility and processability. In addition, the topological stiffness, induced by the aromatic sequence, results in the preservation of high values of electro-optical coefficients during long periods of use [113]. In this way, by coupling reaction, was synthesized the poly(phenylene- imide thiophene) materials (PPIT) (**Figure 10**) [114]. They showed high Tg and consequently high temporal stability of the SHG, as demonstrated by dynamic signal decay behaviour studies that attest their stability at temperatures up to 150°C, preserved more than 82% of the SHG signal after 1500 h of operation.

An elegant and efficient method of synthesis of PImss functionalized with second-order NLO, is the one-step synthesis (**Figure 11**) [115], which allows the use of the most popular electron-acceptor moieties. This give the materials with T_g values in the range 205–224°C [113, 116–119], whose thin films, corona-poled and SHG analyzed, are characterized by enhanced d_{33} value (30 pm / V at λ = 1064 nm). The study of the preservation of SHG properties under thermal stress – as a function of annealing time at 125°C - reveals that the NLO response, after 9–20% initial decrease, does not change significantly over a period of 210 h for polymers: (PIc-e). (PIc), (PId) and (PIe), with better stability than PIb at 125° C, retaining about 70% of the NLO properties (**Figure 11**).

Step-by-step polyaddition reactions is another tool for synthesis of thermally stable, second-order NLO chromophore PIms and poly (urea), starting from

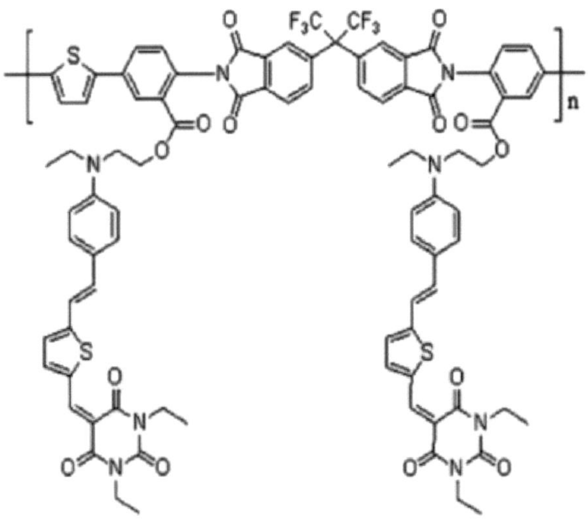

Figure 10.
Structure of new poly(phenyleneimide) (50.85% chromophores, T_g = 170°C; T_d = 245°C, Φ = 23; r_{33} = 35 pm/V).

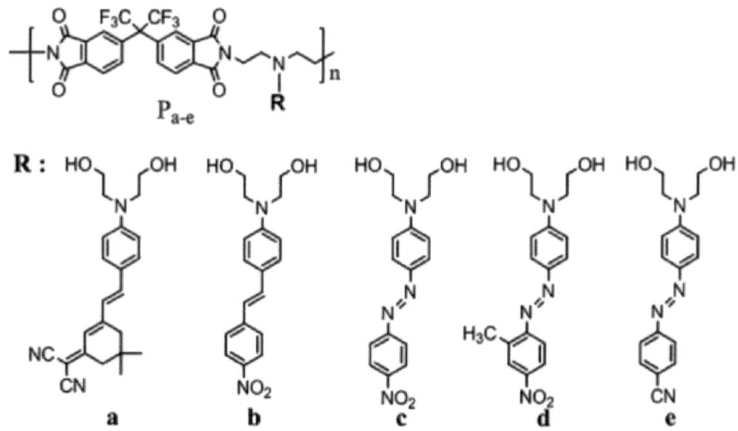

Figure 11.
Poly(imide)s containing NLO chromophores based on 2-(5, 5-dimethyl-cyclohex-2-enylidene)-malonitrile and conventional chromophores.

2,3-bis (4-aminophenyl) -5,6-dicyanopyrazine (BAPDCP), who's the first order molecular hyperpolarizability, β, was evaluated as 123.5 × 10–30 esu [120].

Using a new polymerization methodology, were produced the high T_g polymers [121] introducing into polymer backbone an imide–siloxane linkage containing NLO-active chromophores. As expected, introduction of dimethylsiloxane linkages in the backbone led to enhanced solubility and thermal stability of the functionalized PIms while retaining their useful physical characteristics, such as T_g: These polymers show SHG efficiencies comparable to those for functionalized PIms or polyurethanes and exhibit high temporal stability of resultant SHG signals.

Poly (maleimide) polymers functionalized with aminoalkyl sequences have a transition temperature in the range of 178–228° C, as well as a stable NLO response at high temperatures: such systems lose 24% of their property after 1000 hours at 125° C. The value d_{33} of SHG coefficient, 64.0 pm/V measured at 1064 nm, suggests that these polymers could be useful for NLO-applications [122].

The synthesis of polymers with electro-optical characteristics by sequential self-repetitive reaction (SSRR) [123] offers the possibility of making poly (amide-imide) structures. The promotion, in this case, of the reaction between dysfunctional azo chromophores (DR19, NDPD, DNDA), with an excess of 4,4-methylene-diphenylisocyanate (MDI), ends with the formation of a carbamic structure, useful for obtaining poly (carbamate diphenylisocyanate) (poly CDI- **Figure 12**). Subsequent addition of trimellitic anhydride (TMA) to the poly-CDI solution leads to the poly (N-acylurea) intermediate, characterized by good solubility. The in-situ poling and curing process, of this reactive mixture, favors the formation of the amide-imide structure of the N-acylurea sequence. The correlation of the thermal behaviour with the electro-optical characteristics, for these polymers, highlights values of r_{33}, in the range of 5.2–25.2 pm / V at 830 nm. There is also a proportionality of these values with the chromophore concentration. Such structures are characterized by good thermal stability- (80°C and optical waveguide losses (3.8–6.6 dB/cm at 830 nm) [124], which argues that when using the SSRR technique, the values of the EO coefficients are the consequence of the orientation of the chromophores on the polarization direction before lattice hardening.

Carbazole and its derivatives remain in the attention of chemists, mainly due to the fact that the carbazole skeleton has proven to be a versatile platform for the development of materials with applications in specialized fields such as optics, thermoelectronics as well as medical and pharmaceutical applications.

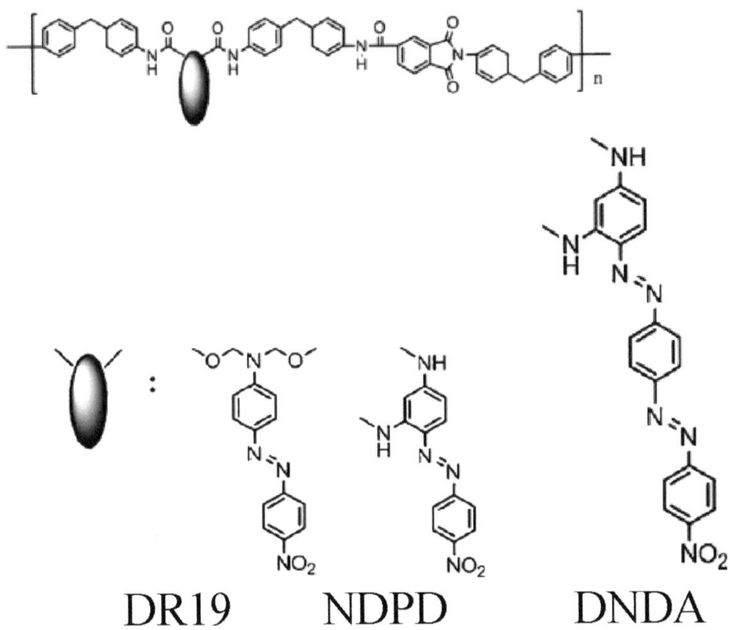

Figure 12.
Poly(amide-imide)s NLO polymers structures obtained by SSRR.

Belonging to a very interesting and relatively new class of material-conjugated polymers, carbazole and its derivatives are manly characterized by the existence of π-mobile electrons, which can be used as chromophores, respectively electrophores [125]. The size of the energy difference between the conduction band and valence band, customizes the wavelength of absorption and / or radiative emissions, configuring the conduction capacity. What differentiates conductive polymers, expressly carbazoles, in the vast class of polymeric materials (known as true insulators) is their ability to carry electrical charges. This behaviour brings them closer to the electrical conductors - metallic or semiconductor.

Such materials are used both in the field of powerful light transmitters for sensors or as active components in electronic (opto) devices and batteries [126–129].

The main advantages of carbazole compounds are related to the economical, processability and their properties, particularly regarding few aspects:

- a cheap raw material readily available from coal–tar distillation, furthermore a large category is retrieved in the plants

- Different substituents can be easily introduced into the carbazole ring the presence of the nitrogen atom in the centre of symmetry, facilitates functionalization with various substituents. Consequently, the properties of the material can be modulated without potentiating the steric interactions with the basic carbazole backbone (**Figure 13**).

- The aromatic configuration gives them high thermal and photochemical stability as well as relatively high mobility of charge carriers;

- Carbazolyl groups easily forms relatively stable radical cations (holes).

Due to the mentioned advantages, a wide range of materials for transporting holes [130, 131], molecular glasses [132–135] or light-emitting materials [136–140]

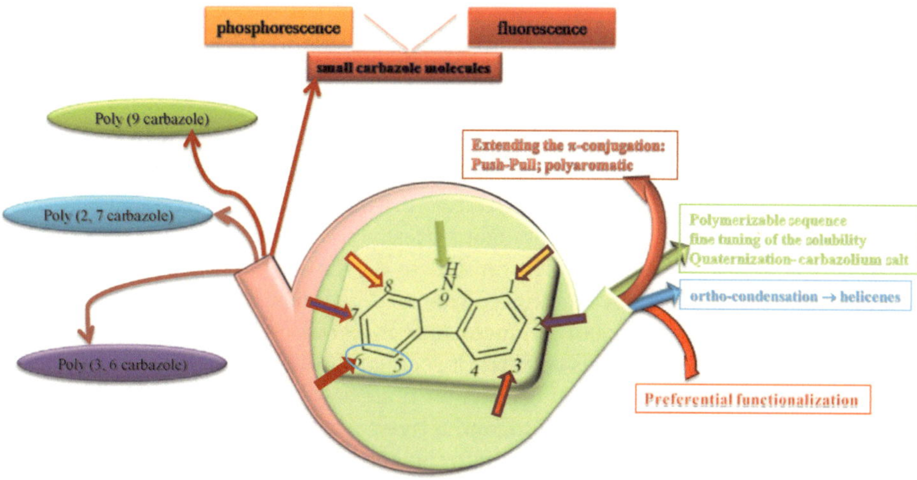

Figure 13.
Carbazole structures: reactive positions correlated with material structure and its particular properties.

have been designed with carbazole. Favourable to the synthesis of different dyes architectures (push-pull dyes; push-push or pull-pull), through substitution reactions to aromatic nuclei, carbazole was and remains an excellent candidate for the design of light harvesting materials for solar cells and numerous dyes with large and wide molar extinction coefficients the absorption spectra were obtained with this plan electron donor [141, 142].

Carbazole derivatives are intensely studied because of their well-known potential as precursors of materials for optical/ optoelectronic applications. The characteristics that recommend them for use are their special photorefractive, electrical, and chemical properties. Carbazoles are well known as a conjugated, good hole-transporting, electron-donor, planar compound and ease to introduce solubilizing groups to rigid ring structure. Carbazole can polymerize and couple in positions 3, 6, 9 and 1, 8 respectively. However, due to the very rigid structure of the carbazole, the last pair of positions is sterically hindered [143, 144]. The very high reactivity of positions 3, 6, favors the rapid synthesis of carbazole derivatives starting from 9H-carbazole, by direct bromination of the carbazole group with N-Bromo succinimide (NBS) [145].

In the context of the progress of new techniques for the manufacture of intelligent devices, the constructive versatility of the carbazole sequence, stimulated the development of design new dyes classes with photoinitiation activity in UV, near-UV or visible light, very useful in the development of 3D printing [146].

Substitution reactions at the carbazole unit can generate poly (3,6-carbazole) and poly (2,7-carbazole), respectively. These structures, due to the effective difference in the length of the conjugate sequence ((2, 7) is the longest, being similar to polyp phenylenes) have different properties and potential applications. In addition, using specialized modification strategies, the properties of both polycarbazole classes can be improved / fine-tuned, thus diversifying high-performance applications in the electronic and electro-optical field, such as polymeric light emitting diodes (PLEDs), organic field-effect transistors (OFETs), and photovoltaic cells (PCs). For both categories of polycarbazoles, an elegant and very complete presentation of the structure-properties-application correlation is summarized in several reviews, among which we recommend [147–149].

From this multitude of papers, the authors summarize studies dealing with novel aspects in poly (3,6-carbazole) synthesis, particular aspects of the photoactive properties of polyvinyl carbazole copolymers, as well as other derivatives with polymerizable group in position 9 of carbazole, adjacent to the poly (2,7 carbazole) derivatives, strongly predisposed for optoelectronic applications such: organic LED, photovoltaic cells and biosensors.

Thus, the first decades of the 21st century are distinguished by the extension of studies dedicated to the synthesis and characterization of new optically active photochromic polymeric structures, with two distinct functional sequences: azo and a chiral group [150–153]. The simultaneous presence of these functions, sensitive to electromagnetic stimuli, creates the premises to display both the typical properties of asymmetric systems (optical activity, exciton division of dichroic absorption) and the typical characteristics of photochromic materials (photo refractivity, photo reactivity, N properties). In this context, it seemed interesting to investigate the properties of new multifunctional copolymer materials containing both photo responsive azobenzene and the photoconductive carbazole chromophore directly bound to the polymeric side chain through the chiral fragment: the presence of the electron-rich carbazole sequence as partner of the electron-poor azobenzene chromophores can induce charge transfer interactions fundamental in the operation of photoconductive materials [154].

The literature highlights the appearance of charge transfer interactions in achiral copolymers of N-vinyl carbazole (NVK) with methacrylic azoderivatives, the intensity of which is stimulated by the alternation of monomer sequences [155–159].

Due to their complexity of structural behaviour, they are able to generate sophisticated synthetic architectures, with high sensitivity and selectivity, which expands the application range. By consequence, they may be adopted special applicative functions, such: with molecules and ions they can be utilized in different separation and purification techniques; by interactions with the electromagnetic stimulus in solar cells fabrication, organic light emitents and optical elements, by specifical interaction with the biological sequences they developed applications in anti-biofouling or specific binding proteins, by specifical interactions in adhesive assemblies, they can generated bonding-disbonding phenomena's, by peculiarity of theirs response to the stimulus action they can behave as self-healing agent for other different materials, by interactions between different polymer's sequences can be achieved a multi-layered polymer materials with different functionalities in applications: screen printing, stamping ink-jet-as commercial used technology or optical waveguide, elastomeric light-emitting devices and displays or molecularly stretchable electronics- as specialized technologies.

4. Conclusions

Exhausting analyzing of the bibliographic information of the NLO materials shows great promise of polymer for device applications in the communications and photonics industries. Usually, these were use as host for chromophore structures as well as statistical (co)polymers, which include the NLO sequence in the macromolecular assembly.

At present there are a large number of NLO applications that use inorganic crystals. A few of these processes include optical switches, EO modulators, optical data storage, and optical fiber-based devices. The study of organic NLO chromophore-doped polymers would allow them to replace their inorganic counterparts. It should be clear to the reader that chromophore design has progressed significantly due to new classes of heterocyclic moieties used to optimize chromophore properties. However, with such novel designs, considerable research is required to achieve highly improved EO activities.

NLO chromophore functionalized polymers (obtained by polymerization or chemical transformation method) were found to be more effective because chromophore mobility is hindered by its attachment to the polymer. Functionalized SC NLO polymer systems with high and stable NLO effects, i.e. $\chi^{(2)}$ and $\chi^{(3)}$ were synthesized. The rigid aromatic backbone approach was found to be promising for making highly stable NLO polymeric systems. Polymers with rigid polymer backbone exhibiting high T_g values were found to suppress reorientation significantly. Theirs exhibiting high thermal stability in dipole orientation, relatively high optical nonlinearity and low optical loss are very promising for practical applications as has been reported.

Few work directions were showed for synthesis of the new stable NLO organic materials:

- Copolymerization provides a general approach to the enhancement of the thermal stability and the NLO stability of the materials within a tolerable trade-off range in optical nonlinearity. Introduction of an imide– siloxane linkage into polymer backbone containing NLO-active chromophores led to enhanced solubility and thermal stability, high glass transition temperatures and exhibit high temporal stabilities of the resultant SHG signals.

- Another strategy for stable polymeric systems is crosslinking technique. Extensive research on this topic for the past few years has shown that the NLO stability is profoundly enhanced by crosslinking the MC.

- More recently, functional dendrimers have been proposed to prepare multi-chromophore material systems that possess unique molecular architecture and characteristics. It is noticeable that a polymer host is not required for certain dendrite materials because they have unique film forming properties due partly to the reasonably high molecular weight and partly to the polymeric character inherent in these molecules. These dendrimer properties allow an appropriate solution viscosity when mixed with traditional spin-casting solvents to create amorphous glassy films. Those demonstrate dramatic enhancements in poling efficiency due to larger free volume and reduced dipolar interactions allowing diminished mobility constraints in comparison to the tethered polymer systems.

- Finally, it should be pointed out that our aim was to underline the main aspects regarding the relation structure- material properties and for that is practically impossible to mention here all investigations. What is not understood for real materials is how medium and chromophore-chromophore interactions modify/can be used to tune, the response of the individual chromophore molecules.

Author details

Ana-Maria Albu and Vlad Marian Târpă*
Department of Bioresources and Polymers Science, Faculty of Applied Chemistry and Materials Science, University Politehnica of Bucharest, Romania

*Address all correspondence to: tarpavlad@yahoo.com

References

[1] Armstrong, J. A., Bloembergen, N., Ducuing, J.; Pershan, P. S, (1962), Interactions between light waves in a nonlinear dielectric, *Phys. Rev.*, 127, 6, 1918-1939. DOI: 10.1103/PhysRev.127.1918

[2] Hurst, M.; Munn, R.W., (1986), Theory of Molecular Opto-Electronics. I. Macroscopic and microscopic response, J. Molec. Electron., 2, 35-41. DOI: 10.1007/978-1-4615-3818-9_20

[3] Boyd, R. W., Nonlinear Optics, Academic Press, San Diego, 1992. DOI: 10.1016/B978-0-444-89304-8.50010-9

[4] Chen, A., Chuyanov, V., Zhang, H., Garner, S., Steier, W. H., Chen, J., Zhu, J., He, M., Mao, S. S. H., & Dalton, L. R., (1998), Low-Vπ electro-optic modulator with a high-μβ chromophore and a constant-biasfield, Opt. Lett., 23, 6, pp. 478-80, ISSN: 0146-9592. DOI: 10.1364/OL.23.000478

[5] Chen, A., Chuyanov, V., Zhang, H., Garner, S., Lee, S.-S, Steier, W. H., Chen, J. Zhu, J., Wang, F., He, M., Ra, Y., Mao, S. Dalton, L. R &Fetterman, H. R., (1998), Low-Vπ high-thermal-stabilityelectro-optic polymer waveguidemodulatorsusingthe full potenetial of high-μβ chromophoreswith a DC biasvoltage, Polymer PhotonicDevices, USA Conference, 28 January 1998, Proc. SPIE 3281, San Jose, CA, Kippelen, B, Bradley D. D. C., Eds, pp. 94-105. The International Society for Optical Engineering. ISSN (printed): 0277-786X. ISSN (electronic): 1996-756X. DOI: 10.1117/12.305409

[6] Dalton, L., Harper, A., Ren, A., Wang, F., Todorova, G., Chen, J., Zhang, C. & Lee, M., (1999), PolymericElectro-optic Modulators: FromChromophore Design toIntegrationwith Semiconductor VeryLarge Scale Integration Electronics and Silica FiberOptics, Ind. Eng. Chem. Res., 38, 1, pp.8-33. DOI: 10.1021/ie9705970

[7] Marder, S. R.; Beratan, D. N.; Tiemann, B. G.; Cheng, L.T.; Tam, W., (1991), Organometallic Materials with Second-order Optical Non-linearities, in Organic Materials for Non-Linear Optics II, Volume 91 of Special publication - Royal Society of Chemistry (Great Britai. Hannand R. A &Bloor D., GhrahamHouse ed., Cambridge, pp. 165-176, ISBN 0851863973, 9780851863979. DOI: 10.1021/bk-1991-0455

[8] Marder, S. R., Beratan, D. N., & Cheng, L.-T., (1991), Approaches for OptimizingtheFirst Electronic Hyperpolarizability of Conjugated Organic Molecules, Science, 252, 5002, pp. 103-106. ISSN: 0036-8075. DOI: 10.1126/science.252.5002.103

[9] Marder, S. R., Gorman, C. B., Tiemann, B. G. &Cheng, L.-T., (1993), Stronger acceptors can diminish nonlinear optical response in simple donor-acceptor polyenes, J. Am. Chem. Soc. 115, 7, pp. 3006-3007. ISSN: 0002-7863. DOI: 10.1021/ja00060a071

[10] Marder, S. R., Perry, J. W., Bourhill, G., Gorman, C. B., Tiemann, B. G., & Mansour, K. (1993), RelationBetween Bond-LengthAlternation and Second Electronic Hyperpolarizability of Conjugated Organic Molecules, Science, 261, 5118, pp. 186-189, ISSN 0036-8075. DOI: 10.1126/science.261.5118.186

[11] Marder, S. R. & Perry, J. W., (1993), Direct observation of reduced bond-lengthalternation in donor/acceptor polyenes, J. Ame. Chem. Soc., 115, 6, pp. 2524-2526. ISSN: 0002-7863. DOI: 10.1021/ja00059a067

[12] Marder, S. R., Gorman, C. B., Meyers, F., Perry, J. W., Bourhill, G., Bredas, J. L., Pierce, B. M., (1994), A

Unified Description of Linear and Nonlinear Polarization in Organic Polymethine Dyes, Science, 265, 51720, pp. 632-635. ISSN: 0036-8075.DOI: 10.1126/science.265.5172.632

[13] Meyers, F., Marder, S. R., Pierce, B M., &Bredas, J. L., (1994), Electric Field Modulated Nonlinear Optical Properties of Donor-Acceptor Polyenes: Sum-Over-States Investigation of the Relationship between Molecular Polarizabilities (α, β and γ) and Bond Length Alternation, J. Am. Chem. Soc., 116, pp 10703–10714, ISSN: 0002-7863. DOI: 10.1021/ja00102a040

[14] Meyers, F., Marder, S. R., Pierce, B M., &Bredas, J. L., (1994), Tuning of large second hyperpolarizabilities in organic conjugated compounds, Chem. Phys. Lett., 228, 1-3, pp. 171-176, ISSN: 0009-2614. DOI: 10.1016/0009-2614 (94)00905-8

[15] Risser, S. M., Beratan, D. N. & Marder, S. R., (1993), Structure-functionrelationships for beta., thefirst molecular hyperpolarizability, J. Am. Chem. Soc., 115, 17, pp. 7719–7728. ISSN: 0002-7863. DOI: 10.1021/ja00070a016

[16] Haller, M., Luo, J., Li, H., Kim, T. D., Liao, Y., Robinson, B. H., et al. (2004), A novel lattice-hardening process to achieve highly efficient and thermally stable nonlinear optical polymers, Macromolecules, 37, 3, pp. 688–690, ISSN: 0024-9297. DOI: 10.1021/ma035393b

[17] Hambir, S. A., Wolfe, D., Blanchardand, G. J. & Baker, G. L., (1997), Adjusting the Third-Order Nonlinear Optical Properties of a Conjugated Polymer Film, J. Am. Chem. Soc., 119, 31, pp 7367–7373, ISSN: 0002-7863. DOI: 10.1021/ja9705194

[18] Marder, S. R.; Torruellas, W. E.; Blanchard-Desce, M.; Ricci, V.; Stegeman, G. I.; Gilmour, S.; Bredas, J. L.; Li, J.; Bublitz, G. U. & Boxer, S. G., (1997), Large Molecular Third-Order Optical Nonlinearities in Polarized Carotenoids Science, 276, 5316, pp. 1233-1236, ISSN 0036-8075 DOI: 10.1126/science.276.5316.1233

[19] Bredas, J. L., Adant, C., Tackx, P. & Persoon, A., (1994), Third-Order Nonlinear Optical Response in Organic Materials: Theoretical and Experimental Aspects, Chem. Rev., 94, 1, pp. 243–278, ISSN: 0009-2665. DOI: 10.1021/cr00025a008

[20] Esser A., Fich, H.; Hadicke, K. H., Paust, J., Schrof, W.&Ticktin, A., (1992), Wavelength Dependent $\chi(3)$ Measurements of Long Chain Carotenoid Derivates, in Organic Materials for Non-Linear Optics III, Volume 137 of Special publication - Royal Society of Chemistry (Great Britain), Ashwell G. J., Bloor D., Ed., Thomas, GhrahamHouse ed., Cambridge, p. 213-217, ISBN: 0851866255, 9780851866253

[21] Nickel, E. G.; Spanker, C. W.; Tang, N.; Hellwarth, R. & Dalton, L. R., (1993), Composites Containing Stabilized Bipolarons in Organic Materials for Non-Linear Optics III, Volume 137 of Special publication - Royal Society of Chemistry (Great Britain), Ashwell G. J., Bloor D., Ed., Thomas, GhrahamHouse ed., Cambridge, pp. 237-240, ISBN: 0851866255, 9780851866253.

[22] Spangler, C. W. & Havelka, K. O., (1991), Stabilization of Polaronic and Bipolaronic States in Electro-active Oligomeric Segments of Non-linear Optic Active Copolymers, in Organic Materials for Non-Linear Optics II, Volume 91 of Special publication - Royal Society of Chemistry (Great Britain),. Hannand R. A, Bloor D., GhrahamHouse ed., Cambridge, pp. 281-288. ISBN 0851863973, 9780851863979.

[23] Natansohn, A. & Rochon, P., (2002), Photoinduced Motions in

Azo-Containing Polymers, Chem. Rev., 102, 11, pp. 4139-4176, ISSN: 0009-2665. DOI: 10.1021/cr970155y

[24] Albert, I. D. L.; Marks, T. J.; Ratner, M. A., (1996), Rational design of molecules with large hyperpolarizabilities. Electric field, solvent polarity, and bond length alternation effects on merocyanine dye linear and nonlinear optical properties, J. Phys. Chem. B, 100, 23, pp.9714–9725 ISSN: 1520-6106. DOI: 10.1021/jp960860v

[25] Castellano, O., Bermudez, Y., Giffard, M., Mabon, G., Cubillan, N., Sylla, M., Nguyen-Phu, X., Hinchliffe, A., Soscun, H., (2005), Ab initio and DFT study of the geometric structures and static dipole (hyper)polarizabilities of aromatic anions, J. Phys. Chem. A; 109, 45, pp.10380–7, ISSN: 1089-5639 DOI: 10.1021/jp051136s.

[26] Jacquemin, D.; Femenias, A., Chermette, H., Andre, J. M. &Perpete, E. A., (2005), Second-order Moller-Plesset evaluation of the bond length alternation of several series of linear oligomers, *J Phys Chem A*, 109, 25, pp. 5734–41. ISSN: 1089-5639 DOI: 10.1021/jp0509111

[27] Sohn, J. E.; Singer, K. D.; Kuzyk, M. G.; Holland, W. R.; Katz, H. E.; Dirk, C. W.; Schilling, M. L. &Comizzoli, R. B.; (1989), Orientationally Ordered Nonlinear Optical Polymer Films, in "Nonlinear Optical Effects in Organic Polymers" NATO Advanced Research Workshop on Nonlinear Optical Effects in Organic Polymers, Sophia-Antipolis, Nice, France 19-20 May 1988, Messier, J., Kajzar, F., Prassad, P., Ulhich, D., Kluwer ed., Acad. Publ., Derdrech, Boston, London, NATO ASI series: Appl. Sc. ser. E, 162, pp. 291-299, ISBN: 0-7923-0132-3. DOI: 10.1007/978-94-009-2295-2_23

[28] Reinhardt, B. A.; Brott, L. L.; Clarson, S. J.; Dillard, A. G.; Bhatt, J. C.; Kannan, R.; Yuan, L.; He, G. S. & Prasad, P. N., (1998), Highly Active Two-Photon Dyes: Design, Synthesis, a nd Characterization toward Application, Chem. Mater., 10, 7, pp. 1863-74, ISSN: 0897-4756. DOI: 10.1021/cm980036e.

[29] Albota, M.; Beljonne, D.; Bredas, J. L.; Ehrlich, J. E.; Fu, J. Y.;Heikal, A. A.; Hess, S. E.; Kogej, T.; Levin, M. D.; Marder, S. R.; McCord-Maughon, D.; Perry, J. W.; Rockel, H.; Rumi, M.; Subramaniam,G.; Webb, W. W.; Wu, X. L. & Xu, C., (1998), Design of Organic Molecules with Large Two-Photon Absorption Cross Sections, Science, 281, 5383, pp. 1653-1656, ISSN 0036-8075. DOI: 10.1126/science.281.5383.1653

[30] Norman, P.; Luo, Y. & Ågren, H., (1998), Structure-to-property relations for two-photon absorption of hydrocarbon oligomers, Chem. Phys. Lett., 296, 1-2, pp. 8-18, :" ISSN: 0009-2614. DOI: 10.1016/S0009-2614(98)01008-2

[31] Norman, P.; Luo, Y. & Ågren, H., (1999), Two-photon absorption in five-membered heteroaromatic oligomers, Opt. Commun., 168, 1-4, pp. 297-303, ISSN: 0030-4018. DOI: 10.1016/S0030-4018(99)00250-3

[32] Norman, P.; Luo, Y. & Agren, H., (1999), Large two-photon absorption cross sections in two-dimensional, charge-transfer, cumulene-containing aromatic molecules J. Chem. Phys., 111, 17, pp. 7758-7766, ISSN: 0021-9606. DOI: 10.1063/1.480111

[33] Porres, L.; Charlot, M.; Entwistle, C. D.; Beeby, A.; Marder, T. B. & Blanchard-Desce, M., (2005), Novelboronquadrupolar NLO-phores: optimization of TPA/transparency trade-off via molecular engineering, Nonlinear Optical Transmission and Multiphoton Processes in Organics III, Proceedings of SPIE Volume: 5934, Yea tes, A. T. Ed., pp. 62-73, ISBN: 9780819459398 DOI: 10.1117/12.618425

[34] Nguyen, K. A.; Rogers, J. E.; Slagle, J. E.; Day, P. N.; Kannan, R.; Tan, L.-S.; Fleitz, P. A. & Pachter, R., (2006), Effects of Conjugation in Length and Dimension on Spectroscopic Properties of Fluorene-Based Chromophores from Experiment and Theory, J. Phys. Chem. A, 110, 49,13172-182, ISSN: 1089-5639. DOI: 10.1021/jp0642645

[35] Ramakrishna, G. & Goodson, T., (2007), Excited-State Deactivation of Branched Two-Photon Absorbing Chromophores: A Femtosecond Transient Absorption Investigation J. Phys. Chem. A, 111, 6, 993-1000, ISSN: 1089-5639. DOI: 10.1021/jp064004n

[36] Yang, Y; Wang, H. et alt.; (2015) The synthesis of new double-donor chromophores with excellent electro-optic activity by introducing modified bridges, *Phys. Chem. Chem. Phys.*, 2015,17, 5776-5784; doi.org/10.1039/C4CP05829A

[37] Lee, S.; Thomas, K. R. J.; Thayumanavan, S. & Bardeen, C. J., 2005, Dependence of the Two-Photon Absorption Cross Section on the Conjugation of the Phenylacetylene Linker in Dipolar Donor–Bridge–Acceptor Chromophores, J. Phys. Chem. A, (2005), 109, 43, pp. 9767-9774, ISSN: 1089-5639. DOI: 10.1021/jp053864l

[38] Barzoukas, M. & Blanchard-Desce, M., (2000), Molecular engineering of push–pull dipolar and quadrupolar molecules for two-photon absorption: A multivalence-bond states approach, J. Chem. Phys., 113, 10, pp. 3951-3959, ISSN: 0021-9606. DOI: 10.1063/1.1288367

[39] Bartkowiak, W.; Zalesny, R. & Leszczynski, J., (2003), Relation between bond-length alternation and two-photon absorption of a push-pull conjugated molecules: a quantum-chemical study, Chem. Phys., 287, 1-2, pp. 103-112, ISSN: 0301-0104. DOI: 10.1016/S0301-0104(02)00982-5

[40] Drobizhev, M.; Karotki, A.; Rebane, A.& Spangler, C. W., (2001), Dendrimer molecules with record large two-photon absorption cross section, Opt. Lett. 26, 14 , pp. 1081-1083, ISSN: 0146-9592. DOI: 10.1364/OL.26.001081

[41] V. Punjabi et al. (Jefferson Lab Hall A Collaboration) 2005, Phys. Rev. C71, 055202, Proton elastic form factor ratios to Q2=3.5GeV2 by polarization transfer DOI: 10.1103/PhysRevC.71.055202

[42] Baur, J. W.; Alexander, M. D. J.; Banach, M.; Denny, L. R.; Reinhardt, B. A.; Vaia, R. A.; Fleitz, P. A. & Kirkpatrick, S. M., (1999), Molecular Environment Effects on Two-Photon-Absorbing Heterocyclic Chromophores, Chem. Mater. 11, 10 , pp. 2899-2906, ISSN: 0897-4756. DOI: 10.1021/cm990258s

[43] Agrawal, G. P. &Flytzanis, C., (1976); Delocalization and superalternationeffects in the nonlinear susceptibilities of one-dimensional systems, Chem. Phys. Lett, 44, 2, pp. 366-370, ISSN: 0009-2614. DOI: 10.1016/0009-2614(76)80532-5

[44] Agrawal, G. P., Cojan, C. & Flytzanis, C., (1978), Nonlinear optical properties of one-dimensional semiconductors and conjugated polymers, Phys. Rev. B, 17, 2, pp. 776-789; ISSN: 1098-0121. DOI: 10.1103/PhysRevB.17.776

[45] Beratan, D. N., Onuchic, J. N. & Perry, J. W., (1987), "Non-linear susceptibilities of finite organic polymers, J. Phys. Chem., 91, 11, pp. 2696-2698, ISSN 0022-3654. DOI: 10.1021/j100295a009

[46] Garito, A. F.; Heflin, J. R.; Wong, K. Y. &Zamaki-Khamiri, O., (1989), Enhancement of Non-linear Optical Properties of Conjugated Linear Chain throughLoweredSymmetry, in “Organic materials for non-linear optics: the proceedings of a conference organised

by the Applied Solid State Chemistry Group of the Dalton Division of the Royal Society of Chemistry, Oxford, 29th-30th June 1988, Issue 69 of Special publication Volume 69 of Special publication // Royal Society of Chemistry", editat de R. A. Hann si D. Bloor, Roy. Soc. Chem., London, pp. 16-28, ISBN:0851868061, 9780851868066

[47] Prasad, P. N., (1989), Studies of Ultrafast Third-order Non-linear Optical Processes in Polymer Films, in Organic materials for non-linear optics: the proceedings of a conference organized by the Applied Solid State Chemistry Group of the Dalton Division of the Royal Society of Chemistry, Oxford, 29th-30th June 1988, 69 of Special publication // Royal Society of Chemistry, Hann, R. A. & Bloor, D., Ed., London 1989, pp. 264-278; ISBN: 0851868061, 9780851868066. DOI: 10.1007/978-94-009-2295-2_28

[48] Spangler, C. W.; Hall, T. J.; liu, P. K.; Dalton, L. R.; Polis, D. W. & Sapochak, L. S., (1991), The design of New Copolymers for $\chi(3)$ Applications: Copolymers Incorporating Ladder Subunits, in Organic Materials for Non-Linear Optics II, Volume 91 of Special publication - Royal Society of Chemistry (Great Britain), Hannand, R. A, Bloor, D., GhrahamHouse ed., Cambridge, pp. 260-267, ISBN 0851863973, 9780851863979.

[49] Spangler, C. W.; Liu, P. K.; Hall, T. J.; Polis, D. W.; Sapochak, L. S. & Dalton, L. R., (1992), The design of new copolymers for x(3) applications, Polymer, 33, 18, pp. 3937-3941, ISSN: 0032-3861. DOI: 10.1016/0032-3861(92)90386-B

[50] Nalwa, S. H., (1991), Recent Developments in Ferroelectric Polymers, J. Macromol. Scie., Part C: Polym. Rev., 31, 4, 341-432, ISSN: 1532-1797. DOI: 10.1080/15321799108021957

[51] Kurihara, T.; Tomaru, S.; Mori, Y.; Shuto, Y.; Hikita, M.; & Kaino, T., (1993), A New Main-chain Polymer for All-optical Waveguide Devices, in Organic Materials for Non-Linear Optics III, Volume 137 of Special publication - Royal Society of Chemistry, Issue 137 of Special publication, Royal Society of Chemistry (Great Britain), Ashwell, G. J., Bloor D., Ed., Thomas Graham Hause, ed., Cambridge, pp. 344-448; ISBN: 0851866255, 9780851866253.

[52] Foss, R. P.; Tam, W.& Zumsteg, F. C., (1991); Preparation of Polymeric Films for NLO Applications, Papers presented at the New York, NY Meeting: Polymeric Materials for Photonic and Optical Applications, 4th Chemical Congress of North America and 202nd ACS National Meeting Published by The Division of Polymer Chemistry, Inc., American Chemical Society, Polym. Prepr. (Am. Chem. Soc, Div. Polym. Chem.), 32(2), 76-78. ISSN: 0032-3934.

[53] Meyrueix, R.; Mignani, G. & Tapolsky, G., (1991), Study of a Novel class of Second Order Non-linear Optical (NLO) Polyurethanes, in Organic Materials for Non-Linear Optics, II, 91, ,. Hann, R. A & Bloor, D. Ed., Roy. Soc; Chem., London, ed., pp. 273-280, ISSN: 0851863973, 9780851863979 DOI: 10.1007/978-94-011-3370-8_11

[54] Kitipichai, P.; la Peruta, R., Korenowski, G. M. & Winek, G. E., (1993), In-situ poling and synthesis of NLO chromophore-bearing polyurethanes for second harmonic generation, J. Polym. Sci., A: Polym. Chem., 31, 6, pp. 1365-1375, ISSN: 1089-5639. DOI: 10.1002/pola.1993.080310603

[55] Eich, M.; Reck, B.; Yoon, Y.; Willson, C. G., Bjorklund, G. C.; (1989), Novel second order nonlinear optical polymers via chemical cross-linking induced vitrification under electric

field, J. Appl. Phys., 66, 7, pp. 3241-3247, ISSN 0021-8979. DOI: 10.1063/1.344115

[56] Hann, R. A., (1989), "Organic Materials for Non-linear Optics", Roy. Soc. Chem., London, Roy. Soc. Chem., ISBN: 0851868061, 9780851868066, London. DOI: 10.1002/adma.19890011214

[57] Prasad, P. N. & Reinhard, B. A., (1990); Is there a role for organic materials chemistry in nonlinear optics and photonics?, Chem. Mater., 2, 6, 660-669, ISSN: 0897-4756. DOI: 10.1021/cm00012a014

[58] Jungbauer, D.; Reck, B.; Twieg, R. J.; Yoon, D. Y.; Willson, C. G. &Swalen, J. D., (1990), Highly efficient and stable nonlinear optical polymers via chemical cross-linking under electric field, Appl. Phys. Lett., 56, 26, pp. 2610-2616, ISSN: 0003-6951. DOI: 10.1063/1.102853

[59] Jungbauer, D.; Teraoka, I.; Yoon, D. Y.; Reck, B.; Swalen, J. D.; Twieg, R. J. & Willson, C. G., (1991); Second order nonlinear optical properties and relaxation characteristics of poled linear epoxy polymers withtolanechromophores, J. Appl. Phys., 69, 12, pp. 8011-8017, ISSN 0021-8979. DOI: 10.1063/1.347497

[60] Reck, B.; Eich, M.; Jungbauer, D.; Twieg, R. J.; Willson, C. G.; Yoon, D. Y. & Bjorklund, C. G., (1990), Crosslinked Epoxy Polymers with Large and Stable Nonlinear Optical Susceptibilities in Proc. SPIE 1147: Proceedings of SpieNonlinearopticalproperties of organic materials II: 10-11 August 1989, San Diego, California, 74-83, G. Khana rian; Ed., Society of Photo Optical ed., ISBN: 0819401838, 9780819401830. DOI: 10.1117/12.962109

[61] Swalen, J. D.; Bjorklund, G. C.; Fleming, W.; Herminghaus, S.; Jungbauer, D.; Jurich, M.; Moerner, W. E.; Reck, B.; Smith, B. A.; Twieng, R.; Willson, C. G. &Zentel, R., (1991), Poled Epoxy Polymers for Optoelectronics, in Organic molecules for nonlinear optics and photonics, 194, NATO Advanced Study Institutes series. Series E, Applied sciences, J. Messier, F. Kajzar, P. N. Prasad, Ed. , Kluwer Academic Publishers, ed. Jluwer Acad. publ., Netherlands, 433-477, ISBN: 0792311817, 9780792311812. DOI: 10.1007/978-94-011-3370-8_31

[62] J. Nordmann, P. Rohl, Cross-linked polymers for non-linear opticswithimprovedlong-termostability, Organic Materials for Non-linear Optics III, Eds. (1993), pp.190–196.

[63] Zentel, R.; Jungbauer, D.; Twieg, R. J.; Yoon, Do Y. & Grant Willson, C.; Synthesis and non-linear opticalcharacte ristics of crosslinked and linear epoxy polymers with pendant tolanechromoph ores. Die Makromol. Chem. Macromol., 1993, 194, 859-868. DOI: 10.1002/macp.1993.021940310

[64] Frinkelman, H.; Ringsdorf, H. & Wendorff, J. H., (1978), Model considerations and examples of enantiotropic liquid crystalline polymers. Polyreactions in orderedsystems, Makromol. Chem., 179, 1 pp. 273-276, ISSN: 1521-3935 DOI: 10.1002/macp.1978.021790129

[65] Ashwell, G. J.; Dawnay, E. J. C.; Kuczynski, A. P. & Martin, P. J., Highe stobservedsecondharmonicintensityf rom a multilayered Langmuir-Blodgett f ilm structure, in "Physical Concepts of Materials for Novel Optoelectronic Device Applications", Physical Concepts of Materials for Novel Optoelectronic Device Applications I: Materials Growth and Characterization Aachen, Federal Republic of Germany, Sunday 28 October 1990, SPIE Proc. 1361, ManijehRazeghi, Ed., pp. 589-598 ISBN: 9780819404220. DOI: 10.1117/12.24423

[66] Stone, P. J. W.; Miller, L. S.; Walton, D. J. & Sethi, R S., (1993),

Langmuir-Blodgett Films of Preformed Polymers for Second Harmonic Generation, in "Organic Materials for Non-linear Optics III", Ashwell, G. J. & Bloor, D., Eds., Royal Society of Chemistry (Great Britain) Special publication; no.137 , Thomas Graham Hause, ed., Cambridge, pp. 88-96; ISBN: 0851866255, 9780851866253.

[67] Rau, H., (1983), Asymmetric photochemistry in solution, Chem. Rev., 83, 5,535–547. ISSN: 0009-2665. DOI: 10.1021/cr00057a003

[68] Rau, H. In Photochemistry and Photophysics; J. K. Rabek, Ed; CRC Press: Boca Raton, FL, (1990); Vol. 2 (chp. 4), p 119-141.

[69] Kumar G. S., (1992), Azo functional polymers: functional group approach in macromolecular design. Lancaster: Technomic Publication; ISBN: 0877629366, 9780877629368

[70] Todorov, T.; Nikolova, L.; Tomova, N., (1984), Polarization holography. 1: A new high-efficiency organic material with reversible photoinduced birefringence Appl. Opt., 23, 23, pp. 4309-4312, ISSN: 1559-128X. DOI: 10.1364/AO.23.004309

[71] Eich, M.; Wendorff, J. Makromol. Chem. (1987), M. Eich and J. H. Wendorff, "Erasable holograms in polymeric liquid crystals," Makromol. Chem. Rappid Comm., 8, 9 pp. 467-471, ISSN: 1521-3927. DOI: 10.1002/marc.1987.030080909

[72] Eich, M.; Wendorff, J.; Reck, B.; Ringsdorf, H., (1987) , Reversive digital and holographic optical storage in polymer liquidcrystals, Makromol. Chem., 8, 1, pp. 59-63, ISSN: 1521-3927. DOI: 10.1117/12.939643

[73] Ichimura, K.; Suzuki, Y.; Seki, T.; Hosoki, A. & Aoki, K., (1988), Reversible change in alignment mode of nematic liquid crystals regulated photochemically by command surfaces modified with an azobenzene monolayer, Langmuir, 4, 5, pp.1214-1216, ISSN: 0743-7463. DOI: 10.1021/la00083a030

[74] Xie, S.; Natansohn, A. & Rochon, P., (1993), Recent developments in aromatic azo polymers research, Chem. Mater., 5, 4, pp. 403-411, ISSN: 0897-4756. DOI: 10.1021/cm00028a003

[75] Ichimura, K.; Seki, T.; Kawanishi, Y.; Suzuki, Y.; Sakuragi, M.; Tamaki, T., (1994), Photocontrol of liquid crystal alignment by command surfaces, In Photoreactive Materials for Ultrahigh-Density Optical Memory: MITI research and development program on basic technologies for future industries; M. Irie, Ed.; Elsevier: Amsterdam, pp. 55-75, ISBN-10: 044481936; ISBN-13: 978-0444819369.

[76] Ichimura, K., (1996), Photoregulation of Liquid Crystal Alignament by Photochromic Molecules and Polymeric Thin Films, In Polymers as Electrooptical Components Macromolecular systems, materials approach Macromolecular Systems Series ResearchReportsEsprit, Shibaev, V. P., Ed.; Springer-Verlag: New York, pp. 138-158; ISBN: 3540594868, 9783540594864. DOI: 10.1007/978-3-642-79861-0_4

[77] Natansohn, A. & Rochon, P., (2001), 2000 Macromolecular Science and Engineering Award LectureThe versatility of azobenzene polymers, Ca nadian Journal of ChemistryVolume 79, 1093-1100;doi: 10.1139/v01-098

[78] Ichimura, K., (2000), Photoalignment of Liquid-Crystal Systems, Chem. Rev., 100, 5, pp. 1847-1874; ISSN: 0009-2665. DOI: 10.1021/cr980079e

[79] Delaire, J. A.; Nakatani, K., (2000), Linear and Nonlinear Optical Properties of Photochromic Molecules and Materials Chem. Rev., 100, 5,

pp. 1817-1846, Print Edition ISSN: 0009-2665; Web Edition ISSN: 1520-6890.DOI: 10.1021/cr980078m

[80] Sutherland R. L., Handbook of nonlinear optics, New York: Marcel Dekker; (1996), ISBN: 0824794265, 9780824794262.

[81] Demus, D., Godby, J., Gray, G.W., Spiess, H.-W., Vill, V., Eds., 1998, Handbook of Liquid Crystals, vol. 3, WILEY-VCH.DOI:10.1002/9783527620760

[82] Sekkat, Z. &Knoll, W., (2002), Chapter II 4: Photoisomerization and Photo-Orientation of Azo Dyes in Films of Polymer: Molecular Interaction, Free Volume, and Polymer Structural Effects, in Photoreactive organic thin films, pp. 108-144 Amsterdam: Academic Press; ISBN: 0-12-635490-1 DOI: 10.1016/B978-012635490-4/50005-6

[83] Tsutsumi, N., Matsumoto, O., Sakai, W. &Kiyotsukuri T., (1996), Nonlinear optical polymers. 2. Novel NLO linear polyurethane with dipole moments aligned transverse to the main backbone, Macromolecules; 29, 2 (January 15, 1996), 592–597, ISSN: 0024-9297. DOI: 10.1021/ma951077o

[84] Tsutsumi, N., Morishima, M. & Sakai, W., (1998), Nonlinear optical (NLO) polymers. 3. NLO polyimide with dipole moments aligned transverse to the imide linkage, Macromolecules; 31, 22, 1998), pp. 7764–7769, ISSN: 0024-9297. DOI: 10.1021/ma9803436

[85] Tirelli, N., Solaro, R. A., Ciardelli, F., Follonier, S., Bosshard, C. H. & Günter P., (2000), Structure–activity relationship of new NLO organic materials based on push–pull azodyes: 4. Side chain polymers., Polymer; 41, 2, pp. 415–421, ISSN: 0032-3861. DOI: 10.1016/S0032-3861(99)00202-5

[86] Liu, Y., Jiang, A., Xiang, L., Gao, J. & Huang, D., (2000) Nonlinear optical chromophores with good transparency and high thermal stability, Dyes & Pigments, 45, 3, 189–193, ISSN: 0143-7208. DOI: 10.1016/S0143-7208(00)00018-8

[87] Luh, T. Y., Chen, R. M., Hwu, T. Y., Basu, S., Shiau, C. W., Lin, W. Y., Jin, B. Y. & Hsu, C. C., (2001), Rational design of polymers for optoelectronic interests, Pure Appl Chem, 73, 2, pp. 243–246, ISSN: 0033-4545. DOI:10.1351/pac200173020243

[88] Takase, H., Natansohn, A. & Rochon, P., (2001), Effect of crosslinking on the photoinduced orientation of azo groups in thin polymer films, J. Polym. Sci., Part B: Polym. Phys., 39, 14, 1686-1696 (2001), ISSN: 109-0488. DOI: 10.1002/polb.1140

[89] Li, Z., Li, J., Qin, J., & Ye, C., (2003), Second-order nonlinear optical property of polysiloxane containing indole-based multifunctional chromophore, Synth. Met., 135-136, pp. 467–468, ISSN: 0379-6779. DOI: 10.1016/S0379-6779(02)00692-6

[90] Natansohn, A., Rochon, P., Gosselin, J. & Xie, S., (1992), Azo polymers for reversible optical storage. 1. Poly[4′-[[2-(acryloyloxy)ethyl] ethylamino]-4-nitroazobenzene] Macromolecules, 25, 8, pp. 2268-2273., ISSN: 0024-9297. DOI: 10.1021/ma00034a031

[91] Cojocariu, C., Rochon, P., (2004), Light-induced motions in azobenzene-containing polymers, Pure Appl. Chem., 76, 7-8, 1479-1497. ISSN: 0033-4545. DOI: 10.1351/pac200476071479

[92] Meng, X., Natansohn, A., Barrett, C. & Rochon, P. (1996), Azo Polymers for Reversible Optical Storage. 10. Cooperative Motion of Polar Side Groups in Amorphous Polymers, Macrom., 29, 3, pp. 946-952, ISSN: 0024-9297. DOI: 10.1021/ma951255d

[93] L. Angiolini; T. Benelli; L. Giorgini; F. Mauriello; E. Salatelli; R. Bozio; A. Daurù; D. Pedron, *Synthesis, chiroptical properties and photoinduced birefringence of optically active methacrylic copolymers bearing side-chain bisazoaromatic moietie s*, Eur. Polym. J.; 2007, 43, pp. 3550 – 3561; doi.org/10.1016/j.e urpolymj.2007.05.020

[94] Angiolini, L., Caretti, D., Giorgini, L., Salatelli, E., (2000), Optically active methacrylic polymers bearing side-chain conjugated azoaromatic chromophores, Synth. Met., 115, 1-3 , pp. 235-239. ISSN: 0379-6779 DOI: 10.1016/S0379-6779(00)00341-6

[95] Angiolini, L., Caretti, D., Giorgini, L., et al., (2000), Optically active polymethacrylates with side-chain l-lactic acid residues connected to push–pull azobenzene chromophores, Polymer, 41, 13, pp. 4767-4780. ISSN: 0032-3861. DOI: 10.1016/S0032-3861(99)00701-6.

[96] Buffeteau, T., Labarthet, F. L., Pezolet, M., Sourisseau, C., (1998), Photoinduced Orientation of Azobenzene Chromophores in Amorphous Polymers As Studied by Real-Time Visible and FTIR Spectroscopies, Macromolecules, 31, 21, pp. 7312-7320, ISSN: 0024-9297. DOI: 10.1021/ma980843z

[97] Poncin-Epaillard, F., Beunet, J., Bulou, A. &Bardeau, J. F., (2001), Synthesis of a new polymer surface bearing grafted azo polymer chains, J. Polym. Sci., Part A: Polym. Chem., 39, 18, pp. 3052-3061, ISSN: 1060-1325. DOI: 10.1002/pola.1286

[98] Wu, Y., Zhang, Q., Kanazawa, A., (1999), Photoinduced Alignment of Polymer Liquid Crystals Containing Azobenzene Moieties in the Side Chain. 5. Effect of the Azo Contents on Alignment Behavior and Enhanced Response, Macromolecules, 32, 12, pp. 3951-3956, ISSN: 0024-9297. DOI: 10.1021/ma990197j

[99] Bellebia, S.; Mesli, A.; Florescu (Albu), A.-M.; Chatzopoulos, F. & Montheard, J. P., (1995), "Nouveaux Colorants pour L'optique Non LinéairePréparés a Partir de Bases schiffDérives du Chlorométhylstyrène", J. Soc. Alger. Chim., 5, 2, pp. 1-9; ISSN: 0024-9297.

[100] Albu, A.-M.; Mărculescu, B.; Boborodea, A. G. & Vasilescu, D. S., (1999), "Synthesis and Characterisation of some Polymers with Applications in Non-linear Optics: I. Polymerisation of Styrene and Methyl methacrylate in the Presence of Coloured Comonomers", E. Polym. J., 35, 12, pp. 2197-2202, ISSN: 0014-3057. DOI: 10.1016/S0014-3057 (99)00018-X

[101] Albu, A.-M.; Mărculescu, B. & Vasilescu, D. S., (1999), "Synthesis and Characterization of Some Polymers with Applications in Non-linear Optics: II. Copolymerisation of Styrene with some Monomers containing Azo-Dyes", E. Polym. J., 35, 12, pp. 2203-2205, ISSN: 0014-3057. DOI: 10.1016/S0014-3057 (99)00019-1

[102] Albu, A.-M., Marculescu, B.; Manaila- Maximean, D. &Trouillet, A., (2000), Preparation of a copolymer for second-order nonlinear optics, SIOEL '99: Sixth Symposium on Optoelectronics, Proc SPIE—Int Soc Opt Engen, 4068 ,Necsoiu T. ; Robu, M. & Dumitras, D. C., Ed. pp. 59–64; ISBN: 9780819437051. DOI: 10.1117/12.378725

[103] Sylla, M.; Mănăila-Maximean, D.; Albu, A. M. & Delaunay, J., (2000), "Synthesis, and Non-linear Optical Properties of New Polymers Based on Coloured Monomers", Polymer, 41, 10, pp. 3507-3511, ISSN 0032-3861. DOI: 10.1016/S0032-3861(99)00610-2

[104] Albu, A.-M., Marculescu, B.; Manaila- Maximean, D. &Trouillet, A., (2000), Preparation of a copolymer for second-order nonlinear optics, SIOEL '99: Sixth Symposium on

Optoelectronics,Proc SPIE—Int Soc Opt Engen, 4068 ,Necsoiu T. ; Robu, M. & Dumitras, D. C., Ed. pp. 59–64; ISBN: 9780819437051. DOI:10.1117/12.378725

[105] Albu, A.-M.; (2009), "Second and Third Nonlinear Optical Response in Azoic Pendant Macromolecular Structures", Rev. Chim., 60, 1, pp. 38-41; ISSN:.

[106] Hvilsted S, Andruzzi F, Ramanujam P. S., (1992), Side-chain liquid-crystalline polyesters for optical information storage, Opt Lett. 17, 17, pp. 1234-1236, ISSN: 0146-9592. DOI: 10.1364/OL.17.001234

[107] Hvilsted S, Andruzzi F, Kulinna C, Siesler, H. W., Ramanujam, P. S., (1995) Novel Side-Chain Liquid Crystalline Polyester Architecture for Reversible Optical Storage, Macromolecules, 28, 7, pp. 2172-2183, ISSN: 0024-9297. DOI: 10.1021/ma00111a011

[108] Pedersen M., Hvilsted, S., Holme, N. C. R., Ramanujam, P. S., (1999), Influence of the substituent on azobenzene side-chain polyester optical storage materials, Macromol. Symp., 137, 1, pp. 115-127, ISSN: 1022-1360; 1521-3900. DOI: 10.1002/masy.19991370112

[109] Andruzzi, L., Altomare, A., Ciardelli, F., Solaro, R., Hvilsted, S., Ramanujam, P. S., (1999), Holographic Gratings in Azobenzene Side-Chain Polymethacrylates, Macromolecules, 32, 2 , pp. 448-454, ISSN: 0024-9297. DOI: 10.1021/ma980160j

[110] Nicolescu, F. A., Jerca, V. V., Albu A.-M., Vuluga , D. M. & Draghici, C., (2008), “Synthesis and Copolymerization Study of New Polyimide Precursors with Potential Application in Optical and Photonic Field”, Mol. Cryst. Liq. Cryst., 486, 1, 38–49; ISSN: 1542-1406; 1563-5287. DOI: 10.1080/15421400801917338

[111] Nicolescu, F. A.; Jerca; V. V., Albu, A.- M.; Vuluga, D. M. & Vasilescu, D. S., (2009), Poly(maleic-amic) Structures for Potential Nonlinear Optical Applications, U.P.B. Sci. Bull., Series B, 71, 1 pp. 75-82, ISSN: 1454-2331.

[112] Nicolescu, F. A.; Jerca, V. V; Albu, A.- M; Vasilescu, D. S. &Vuluga, M. D., (2009), “Synthesis and characterization of organic-inorganic polymers from new methacrylate monomers and silane derivatives”, Optical Materials in Defence Systems Technology VI, 31 August 2009, Berlin, Germany, Proc. SPIE 7487, 748701 (octomber, 2009), Grote, J. G.; Kajzar, F.& Zamboni, R., Ed, 74870N DOI: 10.1117/12.832282

[113] Chen T, Jen A. K. Y., Cai Y., (1996), Two-step synthesis of side-chain aromatic polyimides for second-order nonlinear optics, Macromolecules; 29, 2, pp. 535–539, ISSN: 0024-9297. DOI: 10.1021/ma9512566

[114] Bao Z, Chan WK, Yu L., (1995); Exploration of the stille coupling reaction for the synthesis of functional polymers, J Am. Chem. Soc., 117, 50, 12426–12435, ISSN: 0002-7863. DOI: 10.1021/ja00155a007

[115] Gubbelmans, E.; Verbiest, T.; Van Beylen, M.; Persoons, A.; Samyn, C., (2002), Chromophore-functionalisedpolymides with high-poling stabilities of the nonlinear optical effect at elevated temperature. Polymer, 43, 5, pp. 1581–1585, ISSN: 0032-3861. DOI: 10.1016/S0032-3861(01)00678-4

[116] Yu, D.; Gharavi, A. & Yu, L., (1995), A generic approach to functionalizing aromatic polyimides for second-order nonlinear optics, Macromolecules; 28, 3, pp. 784–786, ISSN: 0024-9297, 1520-5835. DOI: 10.1021/ma00107a017

[117] Yu D.; Gharavi, A. & Yu, L., (1996), Highly stable copolyimides for second-order nonlinear optics.

Macromolecules; 29, 19, pp.6139–6142, ISSN: 0024-9297. DOI: 10.1021/ma951511m

[118] Van den Broeck, K.; Verbiest, T.; Van Beylen, M.; Persoons, A. & Samyn, C., (1999), Synthesis and nonlinear optical properties of high glass transition polyimides, Macromol Chem Phys., 200, 12, pp.2629–2635, ISSN: 1521-3925. DOI: 10.1002/(SICI)1521-3935(19991201)200:12<2629::AID-MACP2629>3.0.CO;2-K

[119] Van den Broeck, K.; Verbiest, T., Degryse, J.; Van Beylen, M.; Persoons, A. & Samyn, C., (2001), High glass transition chromophore functionalised polyimides for second-order nonlinear optical applications, Polymer, 42, pp.3315–3322, ISSN: 0032-3861. DOI: 10.1016/S0032-3861(00)00761-8

[120] Davey, M. H.; Lee, V. Y. ; Wu, L.-M.; Moylan, C. R.; Volksen, W.; Knoesen, A.; Miller, R. D.; Marks, T. J.; (2000), Ultrahigh-Temperature Polymers for Second-Order Nonlinear Optics. Synthesis and Properties of Robust, Processable, Chromophore-Embedded Polyimides, Chem. Mater., 12,1679-1693, doi: 10.1021/cm990483+

[121] Hongwei J. & Kakkar A. K. , (1998), Soluble high-Tg polymers for second-order nonlinear optics from an unusual mix of imide and siloxane linkages in the backbone, Macromolecules, 31, 13, 4170–4176, ISSN: 0024-9297. DOI: 10.1021/ma980309r

[122] Samyn, C., Verbiest, T., Kesters, E., Van den Broeck, K., Van Beylen, M. & Persoons, A., (2000), High glass transition chromophore functionalized poly(maleimide-styrene)s for second order nonlinear optical applications, Polymer; 41, 16, pp. 6049–6054, ISSN: 0032-3861. DOI: 10.1016/S0032-3861(99)00834-4

[123] Chang, H. L; Lin H. L., Wang, Y. C., Dai S. A., Su W. C. & Jeng, R. J., (2007), Thermally stable NLO poly(amide-imide)s via sequential self-repetitive reaction, Polymer, 48, pp. 2046–2055, ISSN: 0032-3861. DOI: 10.1016/j.polymer.2007.02.010

[124] Clays, K. & Coe, B.J., (2003), Design strategies versus limiting theory for engineering large second-order nonlinear optical polarizabilities in charged organic molecules, Chem Mater;15, 3, pp. 642–648, ISSN: 0897-4756. DOI: 10.1021/cm0212111

[125] Qiu, Z., Hammer, B. A.G.; Müllen, K.; 2020, Conjugated polymers – Problems and promises, Progr. in Polym. Scie., 100, 2020, 101179, ISSN 0079-6700, https://doi.org/10.1016/j.progpolymsci.2019.101179.

[126] Wu W, Bazan GC, Liu B. Conjugated-polymer-amplified sensing, imaging,and therapy. Chem 2017;2:760–90.

[127] McGehee MD, Heeger AJ. Semiconducting (conjugated) polymers asmaterials for solid-state lasers. Adv Mater 2010;12:1655–68.

[128] Grimsdale AC, Chan KL, Martin RE, Jokisz PG, Holmes AB. Synthesis oflight-emitting conjugated polymers for applications in electroluminescentdevices. Chem Rev 2009;109:897–1091.

[129] Cheng YJ, Yang SH, Hsu CS. Synthesis of conjugated polymers for organicsolar cell applications. Chem Rev 2009;109:5868–923.

[130] F. Dumur, Carbazole-based polymers as hosts for solution-processed organic lightemitting diodes: simplicity, efficacy, Org. Electron. 25 (2015) 345–361.

[131] F. Dumur, L. Beouch, S. Peralta, G. Wantz, F. Goubard, D. Gigmes, Solution-processed blue phosphorescent OLEDs with carbazole-based polymeric

host materials, Org. Electron. 25 (2015) 21–30.

[132] G. Puckyte, B. Schmaltz, A. Tomkeviciene, M. Degbia, J.V. Grazulevicius, H. Melhem, J. Bouclé, F. Tran-Van, Carbazole-based molecular glasses for efficient solid-state dye-sensitized solar cells, J. Power Sources 233 (2013) 86–92.

[133] T.-T. Bui, S.K. Shah, M. Abbas, X. Sallenave, *G. Sini*, L. Hirsch, F. Goubard, Carbazole-based molecular glasses as hole-transporting materials in solid state dye-sensitized solar cells, ChemNanoMat 1 (2015) 203–210.

[134] F. Goubard, R. Aîcha, F. Tran-Van, A. Michaleviciute, F. Wünsch, M. Kunst, J. Grazulevicius, B. Ratier, C. Chevrot, Investigation of solid hybrid solar cells based on molecular glasses, Proc. Estonian Acad. Sci. Eng. 12 (2006) 96–110.

[135] T.-T. Bui, L. Beouch, X. Sallenave, F. Goubard, Carbazol-N-yl and diphenylamino end-capped triphenylamine-based molecular glasses: synthesis, thermal, and optical properties, Tetrahedron Lett. 54 (2013) 4277–4280.

[136] D. Liu, D. Li, H. Meng, Y. Wang, L. Wu, Multifunctional applications of triazine/ carbazole hybrid thermally activated delayed fluorescence emitters in organic light emitting diodes, J. Mater. Chem. C 7 (2019) 12470–12481.

[137] X.-D. Zhu, Q.-S. Tian, Q. Zheng, X.-C. Tao, Y. Yuan, Y.-J. Yu, Y. Li, Z.-Q. Jiang, L.-S. Liao, A sky-blue thermally activated delayed fluorescence emitter based on multimodified carbazole donor for efficient organic light-emitting diodes, Org. Electron. 68 (2019) 113–120.

[138] J.-L. Cai, W. Liu, K. Wang, J.-X. Chen, Y.-Z. Shi, M. Zhang, C.-J. Zheng, S.-L. Tao, X.-H. Zhang, Highly efficient thermally activated delayed fluorescence emitter developed by replacing carbazole with 1,3,6,8-tetramethyl-carbazole, Front. Chem. 7 (2019) 17.

[139] T.-T. Bui, F. Goubard, M. Ibrahim-Ouali, D. Gigmes, F. Dumur, Thermally activated delayed fluorescence emitters for deep blue organic light emitting diodes: a review of recent advances, Appl. Sci. 8 (2018) 494.

[140] T.-T. Bui, F. Goubard, M. Ibrahim-Ouali, D. Gigmes, F. Dumur, Recent advances on organic blue thermally activated delayed fluorescence (TADF) emitters for organic light-emitting diodes (OLEDs), Beilstein J. Org. Chem. 14 (2018) 282–308.

[141] K. Karon, M. Lapkowski, Carbazole electrochemistry: a short review, J. Solid State Electrochem. 19 (2015) 2601–2610.

[142] F. Dumur, Carbazole-based polymers as hosts for solution-processed organic lightemitting diodes: simplicity, efficacy, Org. Electron. 25 (2015) 345–361.

[143] Elkhidr HE, Ertekin Z, Udum YA, Pekmez K (2020) Electrosynthesis and characterizations of electrochromic and soluble polymer films based on N-substituted carbazole derivates. Synth Met 260:116253. https://doi.org/10.1016/j.synthmet.2019.116253

[144] Madhavan S, Santhanam KSV (1988) Structurally Modified Polymer of Carbazole with Specif ic Cat lysis in Electrooxidation of Organic Molecules. Mol Cryst Liq Cryst Inc Nonlinear Opt 160:111–119. https://doi.org/10.1080/15421408808083006

[145] Boudreault P-LT, Beaupré S, Leclerc M (2010) Polycarbazoles for plastic electronics. Polym Chem 1:127–136. https://doi.org/10.1039/B9PY00236G

[146] Dumur, F., 2020, Recent advances on carbazole-based photoinitiators of

polymerization, Eur. Polym J., 125, 109503, ISSN 0014-3057, https://doi.org/10.1016/j.eurpolymj.2020.109503

[147] Blouin, N.; Leclerc, M., 2008; Poly(2,7-carbazole)s: Structure–Property Relationships, Accounts of Chemical Research, 41, 1110-1119; DOI: 10.1021/ar800057k;

[148] Georgiades, S. N.; Nicolaou, P. G., 2019, Chapter One - Recent advances in carbazole syntheses, in Advances in Heterocyclic Chemistry, Editor(s): Scriven, E. F.V.; Ramsden, C. A.; Academic Press, 129, 1-88, ISSN 0065-2725, ISBN 9780128174739, https://doi.org/10.1016/bs.aihch.2018.10.001;

[149] Nayana, V., Kandasubramanian, B., (2020) Polycarbazole and its derivatives: progress, synthesis, and applications, J Polym. Res. 27, 285, . https://doi.org/10.1007/s10965-020-02254-7

[150] L. Angiolini, R. Bozio, A. Daurů, L. Giorgini, D. Pedron, G. Turco, Chem. Eur. J. 8, (2002) 4241.

[151] L. Angiolini, T. Benelli, L. Giorgini, A. Painelli, F. Terenziani, Chem. Eur. J. 11, (2005) 6053–6353.

[152] L. Angiolini, T. Benelli, R. Bozio, A. Daurů, L. Giorgini, D. Pedron, E. Salatelli, Eur., Polym. J. 41 (2005) 2045.

[153] L. Angiolini, T. Benelli, R. Bozio, A. Daurů, L. Giorgini, D. Pedron, E. Salatelli, Macromol. 39 (2006) 489.

[154] A. Altomare, F. Ciardelli, L. Mellini, R. Solaro, Macromol. Chem. Phys. 205, (2004) 1611.

[155] L. Angiolini, T. Benelli, L. Giorgini, F. Mauriello, E. Salatelli, Macromol. Chem. Phys. 207 (2006) 1805.

[156] H. Li, R. Termine, L. Angiolini, L. Giorgini, F. Mauriello, A. Golemme, Chem. Mater. 21 (2009) 2403.

[157] L. Angiolini, L. Giorgini, F. Mauriello, P. Rochon, Macromol. Chem. Phys. 210, (2009) 77.

[158] L. Angiolini, D. Caretti, L. Giorgini, E. Salatelli, Macromol. Chem. Phys. 201, (2000) 533.

[159] Luigi Angiolini, Tiziana Benelli, Loris Giorgini, Francesco Mauriello, Elisabetta Salatelli, 2009, Relevant chiroptical and thermal properties in optically active methacrylic copolymers containing carbazole and azoaromatic chromophores in the side-chain, Reactive and Functional Polymers, 69, 898-904, ISSN 1381-5148, https://doi.org/10.1016/j.reactfunctpolym.2009.09.005

Chapter 5

Optical Nonlinearities in Glasses

Helena Cristina Vasconcelos

Abstract

The field of photonics has been the target of constant innovations based on a deep knowledge of the nonlinear optical (NLO) properties of materials and especially on information/data technologies. This chapter compiles some of the main physical aspects needed to understand NLO responses, especially in glasses. Any deviation from the linear correlation between a material's polarization response and the electric component of an applied electromagnetic field is an example of nonlinear optic behavior. Heavy metal oxide and chalcogenide glasses offer the largest nonlinear response. For example, high refractive index and high dispersion glasses fall in the type of non-resonant devices, while the resonant ones comprise metal nanoparticle doped glasses. Metal nanoparticles' doped glasses can be prepared by the sol-gel method. The optical absorption spectrum of Ag-doped silica glass shows the presence of an absorption band of surface Plasmon Resonance due to Ag nanoparticles at 420 nm and Z-scan has been used to study the NLO properties. This chapter contains a brief discussion of the basic principles of nonlinear optics, the review of the nonlinear optical of glass in general, and two separate sections concerning the nonlinear optical effects in the glasses doped with quantum dots and metals, respectively.

Keywords: glass, photonics, nonlinear optical (NLO), Kerr effect (third-order nonlinearity)

1. Introduction

Photonics, a field that aims at the study of generation, manipulation, and detection of light, has become essential in modern life. Photonic devices as all-optical switches and modulators play a key role in worldwide data optical communications or optical computing. Since the invention of lasers in the 1960s, there has been a huge increase in the use of devices that use photons (light) instead of electrons. In 1985, a research group of the Southampton University showed the potential of silica glass fibers doped with Er^{3+} ions for applications in long optical transmission systems, at the wavelength region of 1.55 μm, without the need of electronic repeaters [1]. The invention of the erbium-doped fiber amplifier (EDFA) was a key factor in enabling the transmission of long-distance data through silica fiber. The 1.55 μm optical waveband falls in the low-loss transmission window of silica fiber and the amplification band of EDFA's. Sadly, they are still restricted to amplification in the C and L bands. Therefore, optical fibers using linear near-infrared light transmission are only a small fraction of what can be exploited by extending the operating region to the mid- and far-infrared. In fact, silica optical fibers have a non-negligible attenuation of the emitted signal, so if the range of transparency were extended to longer wavelengths, it would have less attenuation. Hence, transparent glasses in the mid and far-infrared wavelength range are well suited to long-distance communication systems due to the Rayleigh dispersion attenuation coefficient varying with λ^{-4}. Nowadays, almost all data flow, including internet, phone calls, etc., goes through fiber optic transmission lines [2] and the field of communications continues to expand to higher data rates and shorter delays to allow more capacity. The demands of the modern world are looking for high-speed communication and therefore it is expected that an overload of data traffic may occur in the telecommunications window that currently operates in the C and L bands. Therefore, an expansion to a wider bandwidth is required which would facilitate data transmission and new amplification materials are needed beyond EDFA's to provide amplification over the optical fiber. This requires overcoming the limitation of peak water absorption around 1.4 μm. All wave fiber was the first to be designed for optical transmission across the entire telecommunications window from 1.3 μm to 1.67 μm (**Figure 1**) [3]. On the other hand, rare-earth (RE) have low solubility in silica glass which limits the interaction length of active devices based on RE doped silica [4]. Besides, silica has high phonon energy which implies that the RE ions transitions will decay non-radiatively; also exhibit a low nonlinear refractive index and so, nonlinear devices based on silica will require high intensities to operate. Finally, silica has a high transmission loss at wavelengths above 2 μm [3].

The necessary increase in the bandwidth excludes the use of EDFA's, leaving fiber Raman amplifiers as the main devices used for that proposes [5]. In fact, amplifiers based on stimulated Raman scattering and four-wave mixing offer additional advantages over EDFAs [6], operate without the need for doping, and can be used at any spectral region [7]. Moreover, the wavelength of the pump laser can be chosen to give a maximum gain at any wavelength range (S, C, or L-band), and the gain bandwidth is higher than that offered by EDFA's (> 100 nm versus 35 nm), which can be enlarged by an appropriate choice of the material [6]. On the other hand, fiber Raman lasers are excellent options for high-power fiber lasers, mainly because of their high output power and broad gain bandwidth, especially in the near-infrared region.

Although silica is widely used in the near-infrared, it limits the wavelength operating range. To overcome these limitations new glasses for optical device

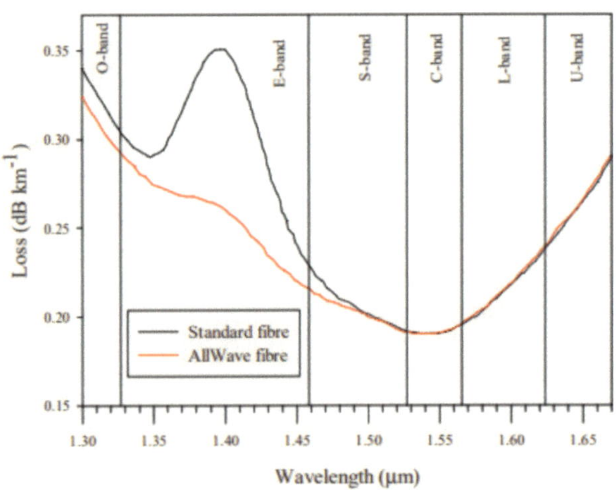

Figure 1.
Loss of standard and all wave silica fibers showing the region of minimum attenuation and the six conventional bands of optical telecommunications [3].

applications and photonics have been investigated. These include heavy metal oxide, fluoride, and chalcogenide glasses.

Glasses containing chalcogenides are the basis for the manufacture of devices operating in the mid-infrared region. In addition, glasses based on heavy metal oxides, such as Sb, Bi, Pb, W, Ga, Ge, Te, allow applications such as optical switches, due to their characteristics of low linear and nonlinear loss, large Kerr nonlinearity, and ultra-fast response. Fluoride-based glasses are used as optical amplifiers in telecommunication as well as in the manufacture of lasers.

Photonics is also used in medical applications, such as lasers used for LASIK surgery, and biomedical diagnostics exploit optical components for bioimaging. Integrated photonics also enables the advance of computing, information technology, sensing, and communications. The integration on a simply planar substrate of several photonic devices (optical sources, beam splitters, couplers, waveguides, detectors, etc.), as proposed by Miller in 1964 [8], enables the control of light on a significantly reduced scale where components are expected to exhibit a very reduced size and achieving a multiplicity of functions, including splitting, combining, switching, amplifying, and modulating signals. Many of these functions are nonlinear. For example, fiber nonlinearities are the basis of several devices such as amplifiers and switching. These nonlinear effects can be divided into two types. The first type is owing to the Kerr-effect (or intensity dependence of the refractive index of the material), which in turn can display phase modulation and wave mixing, depending upon the type of input signal. The second type is related to the inelastic-scattering phenomenon, which can induce stimulating effects such as stimulated Brillouin-Scattering and stimulated Raman-Scattering [9].

NLO is an important issue of advanced photonics and enables technical development in many fields including optical signal processing and quantum optics. It refers to the study of phenomena that occur due to modifications in the optical properties of a material in the presence of light. However, only laser light has sufficient intensity to promote these changes. Indeed, nonlinear optical phenomena (e.g. multiphoton absorption, harmonic generation, self-focusing, self-phase modulation, optical bistability, stimulated Brillouin scattering, and stimulated Raman scattering) require high electromagnetic field intensities to manifest.

2. Basic principles of NLO

In the linear optical domain, photons interact with the glass structure leading to various optical effects, such as dispersion, refraction, reflection, absorption, diffraction, and scattering. For example, the linear refractive index of a material, n, describes how light propagates through it, and the index defines how much light is bent, or refracted when it across the material. However, these properties may become nonlinear if the intensity is high enough to modify the glass optical properties, resulting in the creation of new beam lights of different wavelengths.

A nonlinear optical behavior is a deviation from the linear interaction between a material's polarization response and the electric component of an applied electromagnetic field [10]. This phenomenon involves various optical exchanges such as frequency doubling, conversion, data transformation, etc. Because the magnetic component of light can be ignored in a glass (photons and magnetic fields usually do not interact), the electric component (E) becomes the main field that interacts with the medium. The polarization (P) induced by this interaction produces nonlinear responses that can be explained due to the distortion/deflection of the electronic structure of any atom or molecule (deformation of the electron cloud) due to the application of the electric field, thus producing a resulting dipole moment (vector that separates the positive and negative charges).

Once an external E field is applied to the material the positive charges tend to move in the opposite direction of the electrons. This interaction causes a charge separation that gives rise to microscopic dipole moments within the material. Under the influence of an electric field, these dipoles oscillate at the same frequency (ω) of the incident light. The sum of all the microscopic dipoles of the medium oscillating with time gives rise to material polarization. At low light intensities, Hook's law is valid and the deformation of the electrons cloud is proportional to the applied field strength of the incident light: the light waves and excited electrons oscillate sinusoidally. The induced polarization is also oscillatory and is directly proportional to the incident electric field, as described by:

$$P = \varepsilon_0 \chi^{(1)} E \tag{1}$$

where ε_0 is the vacuum permittivity and $\chi^{(1)}$ (or χ) is the linear susceptibility, which, in this case, depends on the frequency, and thus is directly linked to the linear refractive index, but does not depend on the amplitude of the electric field, which implies that the frequency of light does not change as it passes through matter. However, at high intensities, the electrons are extremely deflected from their orbit, and their movements become distorted giving rise to important deviation from harmonic oscillation. As a result, the amplitude of dipoles oscillation increases, and they emit light not only at the wavelength that excites them but in other frequencies (new color!!!) (**Figure 2**) [11]. At large intensities, P is a nonlinear function of E whereas, at low intensities, the interaction is a linear function. So, for materials with nonlinear characteristics, in which the polarization given by the Eq. (1) is no longer valid, P must be written in a more general form, as a power series of E:

$$P = \varepsilon_0 \left(\chi^{(1)} E^1 + \chi^{(2)} E^2 + \chi^{(3)} E^3 + \dots \right) \tag{2}$$

where the values of $\chi^{(2)}$ and $\chi^{(3)}$, are, respectively, the second-order and third-order susceptibilities which appear due to the nonlinear response of charged particles and are determined by the symmetry properties of the medium. Consequently, nonlinear refractive index (n_2), second ($\chi^{(2)}$), and third-order ($\chi^{(3)}$) nonlinear

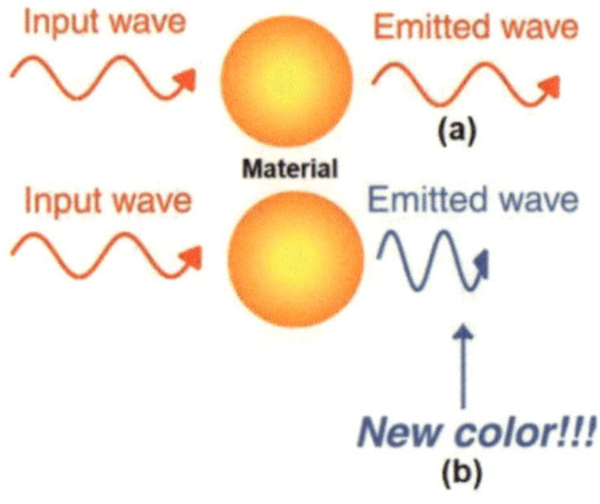

Figure 2.
(a) Linear optics, a light wave acts on the material constituents, which vibrates and then emits its own light wave that interferes with the original light wave, (b) nonlinear optics. Adapted from [11].

susceptibilities can be measured. In isotropic, nondispersive, and homogeneous mediums, the material susceptibilities can be considered constants. However, in anisotropic media where properties are directionally dependent, the susceptibilities of the material are tensor quantities and therefore, depend on the microscopic structure (electronic e nuclear) of the material [12].

Considering the relations $n^2 = 1 + \chi^{(1)}$, $c = \frac{1}{\sqrt{\varepsilon_0 \mu_0}}$, and $n = \frac{c}{c_0}$, where n represents the linear refractive index, c is the speed of light in vacuum, c_0 the speed of light in the material, and μ_0 the vacuum permeability; Maxwell's equations can be used to obtain the wave equations in a nonlinear material:

$$\nabla^2 E - (1 + \chi)\varepsilon_0 \mu_0 \partial_t^2 E = \frac{1}{\varepsilon_0 c^2} \partial_t^2 P \tag{3}$$

where the term $\partial_t^2 P$, represents a measure of the acceleration of the charges that constitute the material, which plays a fundamental role in the theory of nonlinear optics. This term acts as a source in the generation of new radiation field components, producing oscillating electric fields within a linear medium of refractive index n.

Assuming an external electric field of the type $E(t) = E\exp(-i\omega t) + c.c$, where c.c. denotes "complex conjugate", the term related to second order polarization is given by:

$$P^{(2)} = \varepsilon_0 \chi^{(2)} E^2 = 2\varepsilon_0 \chi^{(2)} |E|^2 + \varepsilon_0 \chi^{(2)} E^2 \exp(-2i\omega t) + c.c \tag{4}$$

being responsible for the generation of a field with twice the frequency of the incident radiation (2ω), taking the designation of the second harmonic generation process. However, in centrosymmetric materials, or isotropic materials like glass, which have macroscopic inversion symmetry, the polarization must reverse when the optical electric field is reversed, which implies that $\chi^{(2)}$ must be zero, i.e., all second-order components of the susceptibility tensor are null and GSH does not manifest unless the glass has been poled. It is possible to induce GSH in glasses to break its centrosymmetry, using heat treatments or high energy excitation in the UV [13]. But, without the use of this strategy to eliminate glass's isotropy, only a $\chi^{(3)}$ is $\neq 0$ and may lead to NLO character in glass [10] and the dominant term in (2) is then the third order:

$$P^{(3)} = \varepsilon_0 \chi^{(3)} E^3 \tag{5}$$

which will give rise to frequency tripled light, called third-harmonic generation (THG). According to (5) this nonlinear polarization contains a component of frequency ω and an additional one at 3ω:

$$P^{(3)} = \underbrace{3\varepsilon_0 \chi^{(3)} |E|^2 E}_{P(\omega)} + \underbrace{\varepsilon_0 \chi^{(3)} E}_{P(3\omega)} \tag{6}$$

The term P (3ω) shows that the THG of light is produced while the term P (ω) denotes an incremental change of the susceptibility ($\Delta\chi$) at the frequency ω, given by:

$$\varepsilon_0 \Delta\chi = \frac{P(\omega)}{E} = 3\chi^{(3)} |E|^2 = \frac{6}{n\varepsilon_0 c} \chi^{(3)} I \tag{7}$$

Where I is the intensity of the incident light that become significantly the value of $\chi^{(3)}$. The $\chi^{(3)}$, which gives the dependence of refraction on the intensity of the propagated optical beam, is responsible for the lowest order nonlinear effects in the glass as self-phase modulation and other parametric effects.

Since $n^2 = 1 + \chi$, $\Delta\chi$ is equivalent to an incremental change in the refractive index, Δn is an increase (or decrease) of the total refractive index due to nonlinear effects:

$$\Delta n = \left(\frac{\partial \chi}{\partial n}\right)^{-1} \Delta\chi = \frac{\Delta\chi}{2n} = \frac{3}{n^2\varepsilon_0^2 c}\chi^{(3)} I = n_2 I \tag{8}$$

where n_2 is the nonlinear refractive. This change of the linear refractive index, n, is proportional to the light intensity, and therefore it becomes a linear function of I:

$$n(\mathrm{I}) = n + n_2 I \tag{9}$$

$$n_2 = \frac{3}{n^2\varepsilon_0^2 c}\chi^{(3)} \tag{10}$$

The intensity-dependent refractive index is generally given as:

$$n(\mathrm{I}) = n + n_1 E + n_2 E^2 \tag{11}$$

where n_1 is the Pockel's coefficient (insignificant for isotropic materials as glasses) and n_2 is known as the Kerr coefficient (from the optical Kerr effect) [10]. However, the classical wave theory says that the intensity of the electric field of the light is equal to the square of its amplitude, and thus one can also write n(I) in the form of Eq. (9). The optical Kerr effect is very sensitive to the operating wavelength and polarization dependence and so the prevalent non-linearity occurs at a frequency well below the glass band gap and this effect is called non-resonant [10].

Typical values of the Kerr coefficient (in cm^2/W) are 10^{-16} to 10^{-14} in transparent crystals and glasses. Silica glass (e.g. silica fibers), has an n_2 index of $2.7 \times 10^{-16} cm^2/W$ at the wavelength of 1500 nm, whereas most of the chalcogenide glasses exhibit higher values, about several orders of magnitude larger than silica [14]. Since the values of the nonlinear refractive index in glasses are very small, resulting in a slight change of $\Delta n = n_2 I$, the effect is measurable only for very intense light beams (lasers) of the order of $1 GWcm^{-2}$. From **Figure 3**, it can be noted that n and n_2 are usually directly correlated, such that high index (n) glasses, like chalcogenides, have also high n_2 [16] and exhibit ultrahigh n_2 greater than silica, as plotted in **Figure 3**.

For all-optical signal processing and switching devices, glasses with large n (hence a large n_2) are very attractive. **Figure 4** shows the relationship between the linear refractive index (n), and the third-order nonlinear optical susceptibility $\chi^{(3)}$ of various types of glass. High index (n) glasses, like chalcogenide ones, have also high n_2, which seem to have the largest non-resonant third-order optical nonlinearities related so far. As previously mentioned, $\chi^{(3)}$ arises from light-induced changes in the refraction index that result in the Kerr effect or in parametric interactions (mixing of optical beams). In a glass fiber, the third-order susceptibility is related to n_2 by Eq. (10) and the magnitude of the corresponding nonlinear effect is given by:

$$\gamma = \frac{2\pi}{\lambda A_{eff}} n_2 \tag{12}$$

where λ is the free-space wavelength and A_{eff} is the efficient core area [6]. Since 1999, single-mode silica fibers with γ of 20 W^{-1} km^{-1} were fabricated [18] with a

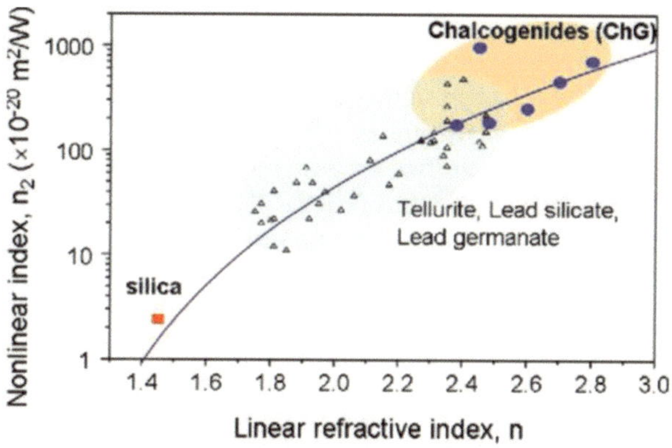

Figure 3.
Nonlinear refractive index, n_2, versus refractive index, n, for various glasses, and silica glasses. Adapted from [15].

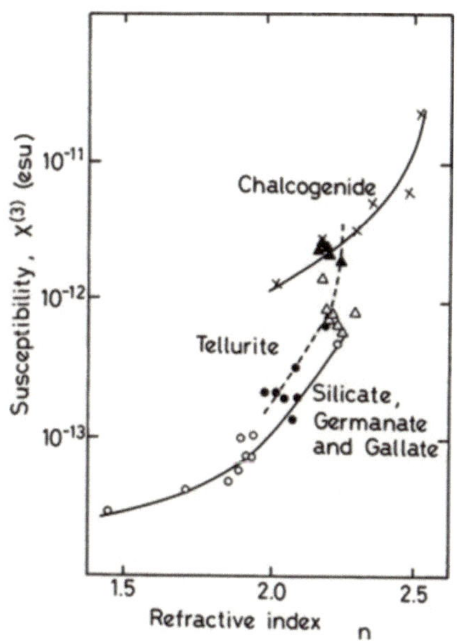

Figure 4.
Relationship between linear refractive index and third-order optical susceptibility. Adapted from [17].

core that was only 10.7 μm^2, but typical A_{eff} values in silica fibers can reach 50 μm^2 for 1.5 μm wavelengths. The self-phase modulation is a phenomenon arising from the dependence between the refractive index of a nonlinear medium and the strength of the electric field, which induces a phase shift of the propagating light, $\varphi_{NL}(z)$:

$$\phi_{NL}(z) = \gamma P_0 z = \frac{z}{L_{NL}} \tag{13}$$

where P_0 is the input power and L_{NL} is the non-linear length that corresponds to the propagation distance at which the phase modulation becomes relevant, being defined by:

$$L_{NL} = (\gamma P_0)^{-1} \tag{14}$$

If the input power is only 1 mW at λ =1.55 μm, and the A_{eff} = 50 μm^2, the L_{NL} is ~500 m [6]. As the refractive index in silica is weakly dependent on power, non-linearities are introduced into the signal propagation and significantly increase in optical networks over relevant distances.

The various nonlinearities can be expressed in terms of the real and imaginary parts of each of the nonlinear susceptibilities $\chi^{(1)}$, $\chi^{(2)}$, $\chi^{(3)}$, ... that appear in (2). The real part is associated with the refractive index and the imaginary part with a time or phase delay in the reply of the material, giving rise to loss or gain. **Table 1** exhibits the principal third-order NLO effects usually showed by dielectric materials like most glasses. For example, the nuclear contribution to stimulated Raman scattering (resulting in loss or gain) can be expressed in terms of the imaginary part of a $\chi^{(3)}$ susceptibility, while the four-wave mixing, which is only of electronic nature and almost an instantaneous effect, result in frequency conversion and in related to the real part of the $\chi^{(3)}$ susceptibility [6]. The imaginary part of $\chi^{(3)}$ provides a change in the absorption coefficient, α, as a function of light intensity:

$$\alpha(I) = \alpha_0 + \beta I \tag{15}$$

where α is the linear absorption, and β is the non-linear absorption coefficient. As a result, occurs a prevalence of non-linearities at frequencies above the electronic absorption edge is known as resonant. The third-order non-linearity may be analyzed in phase conjugate mirrors, like in Mach-Zehnder interferometer pulse selectors or in Fabry-Perot interferometers filled with a nonlinear medium.

The $\chi^{(3)}$ susceptibility is often measured by degenerate four-wave mixing, by the maker fringe method (THG method), or by the Z-scan method. The latter is by far the most used and meticulous method involving the analysis of third-order nonlinear optical properties arising from pulsed laser or CW irradiation at a given wavelength [20].

3. Nonlinear optical properties of glass

Glass is defined as a solid material of amorphous (non-crystalline) structure while crystals possess long-range order, the amorphous materials only possess

Order	Tensor	Effect	Description
3	$\chi^{(3)}(-\omega;\omega,-0,0)$	Kerr's effect	Under the action of two electric fields, there is a change of the refractive index in the NLO medium.
3	$\chi^{(3)}(-\omega;\omega,-\omega,\omega)$	Nonlinear refractive index also called Kerr's effect, self-phase modulation.	The refractive index of the medium changes with intensity according to the formula: $n = n_0 + n_2 I$. Self-focusing and self-defocusing of a laser beam are special cases.
3	$\chi^{(3)}(-3\omega,\omega,\omega,\omega)$	Third harmonic generation.	There is an emission of light with triple frequency under the illumination of the medium.
3	$\chi^{(3)}(-\omega_4;\omega_1,\omega_2,\omega_3)$	Multiwave mixing.	When illuminated with three light sources with different frequencies a generation of light occurs whose frequency equals the sum of the three excitation frequencies.

Table 1.
Third-order NLO effects are usually shown by dielectric materials. Adapted from [19].

short-range order. Therefore, glasses are typically brittle and optically transparent because they lack internal structure. The silica-based glass was undoubtedly the most studied given its multiple applications. Glasses that do not include silica as a main constituent exhibit other properties that make them useful for various applications, for example in optical fibers that work in different frequency domains than SiO_2 fibers. These include fluoride glasses, tellurite glasses, aluminosilicates, phosphate glasses, borate glasses, and chalcogenide glasses. Common glasses are transparent materials in the spectral range of the visible and near-infrared region, although opaque in the far IR and UV region. The visible transparency threshold ends, for high wavelengths (λ), with UV absorption, due to electronic transitions between valence band levels and unfilled conduction band levels. For applications in photonics, there are two main categories of special glasses: chalcogenide glasses (CGs) and heavy metal oxide glasses. Chalcogenide glasses are based on the chalcogen elements S, Se, and Te. These glasses are formed by the addition of other elements such as Ge, As, Sb, Ga, etc. Heavy metal oxide and chalcogenide glasses offer the largest nonlinear response.

Most of the glasses are prepared by the melt of precursors. In solid form, glass is a non-crystalline (or amorphous) material. The deposition from a liquid solution (sol–gel method) is an alternative approach to obtain glass, especially in films form. Some compositions may otherwise be rather difficult to prepare by melt and that's why in practice this method is limited to a relatively small number of compositions. Therefore, the sol–gel processes allow the synthesis of glasses of extended composition ranges, allowing the fabrication of multiple oxide composition, but also non-oxide glasses, with a high degree of homogeneity, because reagents are mixed at the molecular level at temperatures lower than those required for conventional melting. However, the OH content of the sol–gel glasses is high and OH absorptions usually limit transmission at 1.4 μm.

Optical glasses are optically homogeneous glass that are applied in several optical functionalities. The first optical quality (flint) glasses were created at the end of the 19 century by Otto Schott, who also invented Ba crown glass, allowing the production of adjusted lenses for chromatic aberration [21]. X-ray diffraction (XRD) allows distinguishing a glass from a crystalline material. The pattern of SiO_2 glass contains only a few, very broad peaks, which cannot be correlated by the Bragg law with planar distances (as in the case of crystals). SiO_2 consists of a matrix of SiO_4 tetrahedra (**Figure 5**) [22].

The presence of a glass modifier together with the glass formers (SiO_2 or P_2O_5) breaks up the oxide network M–O–M (M = Si, P) and drives the transformation of the bridging oxygens (BO) into nonbridging oxygens (NBOs). The structural unit of SiO_2 has Si-O atomic bonds whose electronic transitions occur in the UV range. For high λ, the transparency threshold ends due to the vibrations of the ions in the network (in resonance with the incident radiation). The amorphous character of the glass explains the absence of grain boundaries in its structure and, therefore, the absence of internal dispersion and reflection phenomena, which are always present in crystalline materials. Glasses are dielectric materials and therefore exhibit a large energy gap between the valence band and the conduction band in accordance with the band theory of solids. Their optical transmission is limited by electronic transitions (Urbach tail) for low wavelength, and multiphonon absorption at high wavelength, in the IR spectrum. The multiphonon absorption process is related to the fundamental vibration frequencies of the glass.

The transmittance spectrum varies from glass to glass, but the main differences are observed outside the transparency range (**Figure 6**). The glass has an optical transparent window which strongly depends on the compositions. Glasses made for use in the visible region have high transmittance across the entire wavelength range

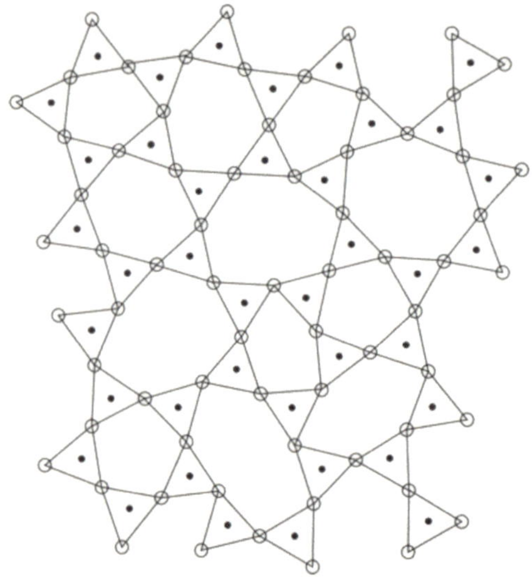

Figure 5.
A schematic representation of the structure of vitreous silica. The tetrahedral SiO4 units in silica are represented by triangular units [22].

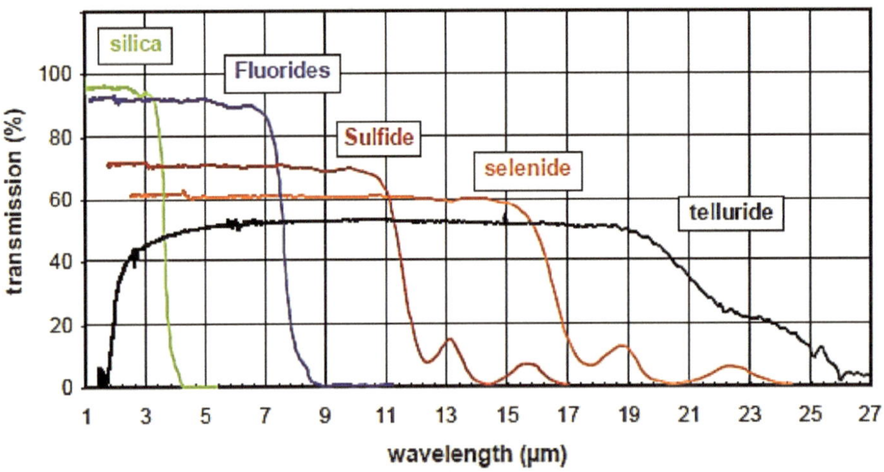

Figure 6.
Typical transmittance spectra of silica, fluorides, sulfide, selenide, and telluride glasses [23].

of ~400 nm–800 nm. However, the structure of silicate glasses limits its transmission in the infrared region to above 3 µm. They have strongly bound electrons but non-bridging oxygens, with their weakly bound electrons, reduce transmission. Chalcogenide glasses, heavy metal fluoride glasses, and heavy metal oxide glasses extend this transmission to higher wavelengths. The telluride glasses have larger atoms and weaker bonds than oxide glasses and so its vibrational resonance occurs at a lower frequency, shifting the fundamental absorption cut-off to longer wavelengths (**Figure 6**).

The interest in chalcogenide glasses backs from 1950s when was reported high infrared transparency of the As_2S glass, up to 12 µm [24]. The structure of chalcogenide glasses such as Ge-Sb-Se consists of covalently bonded atoms, like amorphous SiO2, with lacking periodicity. They include sulfide, selenide, and telluride-based glasses. As dielectric materials, their optical transparent window is dependent on electronic absorption at low wavelengths and multiphonon absorption at high wavelengths. They have a band gap (E_g) that is dependent on the composition and decreases according to Sulfides < Selenides < Tellurides. A specificity of tellurides that differentiates it from the sulfides and selenides in its crystalline structure and physical properties is the large atomic number of Te. The energy gap may be taken from the glass absorption spectrum $\alpha(\hbar\omega)$ by extrapolating the linearized Tauc equation:

$$\alpha\hbar\omega \propto \left(\hbar\omega - E_g\right)^{1/2} \tag{16}$$

The absorption coefficient, α, varies exponentially with the photon energy, $\hbar\omega$ in the Urbach tail.

It is interesting to note that n and n_2 and are usually directly correlated, such that low index (n) glasses, like certain fluorides and phosphates, have also low n_2. On the other hand, a relationship between the material band gap and the n_2 was also established. For example, the n_2 value obtained for pure As_2S_3 was about 2.9 x 10^{-18} m^2/W while for fused SiO_2 was about 2.8 x 10^{-19} m^2/W [25], which is comparatively about 10 times lower. So, materials with lower band gap seam to exhibit an increase in the nonlinear optical behavior; SiO_2 has a gap of about 9 eV while that of As_2S_3 is 2.3 eV [19].

The increase of the nonlinear absorption coefficient (β), third-order nonlinear optical susceptibility ($\chi^{(3)}$), and nonlinear refractive index (n_2) and decreasing the optical band gap (Eg) can be attributed to the formation of BO bonds and ions of higher polarizability in the glass matrix. It has been recognized the effect of the glass composition on the dependency of $\chi^{(3)}$. In most multicomponent oxide glasses, there are both BO and NBO oxygens in the glass network (e.g. for a silicate glass, $Si\text{-}O^+Na^-$). The NBO bonds possess larger n_2 than the BO of the more covalent Si-O-Si bonds [26]. It was also established that third-order nonlinear optical susceptibility of the glasses increases with increasing optical basicity and tendency for metallization of the glasses. This fact is associated with the polarizability of the anions ($F^- < O^{2-} < S^{2-} < Se^{2-}$) and the small optical band gap [19], which is related to the increasing metallicity of the oxides [27]. The theory of metallization of the condensed matter says that in the Lorentz–Lorenz equation, the refractive index becomes infinite when metallization of covalent solid materials occurs [27]. SiO_2, B_2O_3, and GeO_2 based glasses exhibit low refractive index and have low polarizability, large metallization tendency, and small $\chi^{(3)}$. Tellurite and TiO_2 based glasses, as well as B2O3 glasses containing a large amount of Sb2O3 and Bi2O3 with high refractive index, show large polarizability, small metallization tendency, and large $\chi^{(3)}$ (**Figure** 7). Consequently, under the point of view of polarizability, high-refractive-index glasses with an increased tendency for metallization are promising materials for application as components of nonlinear optical devices.

Glass materials are excellent non-linear optical materials, being isotropic and transparent in a wide spectral range, combining low cost of fabrication with high optical quality, manufacturable not only as bulk shapes, or fibers, but also as thin films (e.g. nonlinear planar waveguides). Furthermore, when compared to polymers, glass is more stable and has the advantage over crystals since its atomic composition is easily tailored: a nonlinear optical glass can be obtained with any refractive index in a wide range [28]. Its properties can be adjusted through doping and compositional changes to fit the specified requests of each application. Its disordered structure allows light propagation inside that medium like no other material. They also exhibit good compatibility with silica-based systems and waveguide production in which high optical intensities and long interaction lengths can be achieved [28], giving rise to nonlinear structures in integrated optical devices [29].

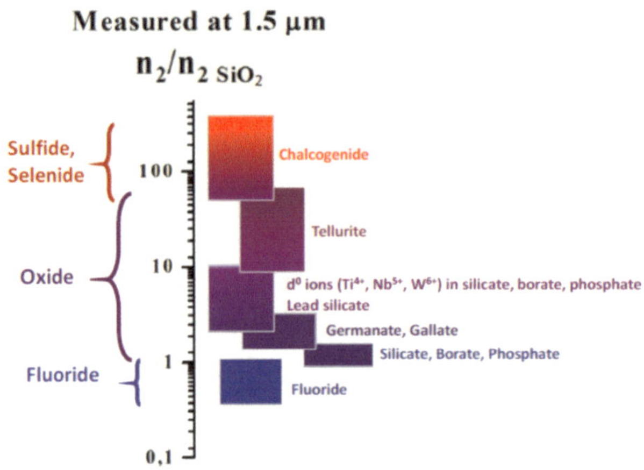

Figure 7.
Line-up of the Kerr effect among various glass compositions [19].

For the fabrication of all-optical systems in information technology and integrated photonics, the chosen materials should exhibit high nonlinearities. Rather, low nonlinearities are essential for fibers in optical communications to avoid phenomena of self-focusing, self-phase modulation, Raman and Brillouin scatterings. NLO was considered the threshold to the total of information that can be transmitted in a single optical fiber. As laser power levels increase, NLO limits data rates, transmission lengths, and the number of wavelengths that can be transmitted simultaneously. Optical nonlinearities give rise to many "secondary" effects in optical fibers. These effects can be damaging in optical communications, but they find other applications, especially for the integration of all-optical functionalities in optical networks. The optical nonlinearities can give rise to gain or amplification, the conversion between wavelengths, the generation of new wavelengths or frequencies, the control of the temporal and spectral shape of pulses, and switching [6]. Thus, they can be distinguished in two types: that from scattering (stimulated Brillouin and stimulated Raman) and that from optically induced changes in the refractive index, resulting either in phase modulation or in the mixing of several waves and the generation of new frequencies (modulation instability and parametric processes, such as four-wave mixing). So, the nonlinear refractive index, also referred optical Kerr nonlinearity (n_2), offers a means to achieve switching and amplifying functions in photonic devices and produces nonlinear effects, namely self-phase modulation, and four-wave mixing. Self-phase modulation implies changes in the phase and rising frequency of a pulse, which can cause spectral broadening. Four-wave mixing is a kind of nonlinear frequency conversion generated by the Kerr nonlinearity which enables, for example, high-speed communications, frequency conversion, sensing, and quantum photonics. The effect of ultrafast response time (10 s^{-15} s) provides broad bandwidths, that can pull actual GHz electronic computing forward to PHz (10^{15}) rates using all-optical signal processing [30]. In addition, spectral broadening, produced by changes in phase from the nonlinear refractive index, can enable the production of short-pulsed sources [30]. Four-wave mixing, on the other hand, can be used to generate optical frequency combs [30], which can measure precise frequencies of light and span spectral ranges useful for spectroscopic investigations.

Although these applications are of great practical interest, the Kerr effect (n_2) is often small for common optical glasses ($\sim 10^{-20}$ to 10^{-19} m^2/W) [30], leading to high thresholds for nonlinear effects and requiring special sources of high-power excitation.

Transparent optical glasses exhibiting nonlinearities, e. g. large nonlinear refractive index and nonlinear absorption coefficient are good candidates for fiber telecommunication and for nonlinear optical devices such as optical switches, self-focusing, and white-light continuum generation. Glasses that exhibit significant nonlinearity are good candidates as Raman gains media to provide enhanced Raman gain over an extended wavelength range. Chalcogenide (As–Se) glasses and fibers are examples of good candidates as well tellurite fibers because of the high refractive index of TeO_2 (2.3–2.4) [6] compared to the SiO_2 (1.46). An As_2S_3 fiber exhibit a Raman coefficient is 300 times greater than that of silica fiber [6]. However, chalcogenide fibers have lesser chemical stability. In spite of that, chalcogenide glass has wide transparency transmission from 0.5 to 25 μm [31], enhancing their potential applications on the mid-IR. As shown in **Figure 8**, the long-wavelength cut-off edges of chalcogenide glasses depend on the mass of anionic elements and are extended between 12 and 20 μm. Their nonlinearity (Kerr effect) is 200–1000 times larger than that of the silica glass at a wavelength of 1.55 μm [32].

The nonlinear optical properties of glasses have been considered of great interest for photonic devices to be used in several technological applications with a broad spectrum of phenomena, such as optical frequency conversion, optical solitons, phase conjugation, and Raman dispersion. Most of the previous investigations were devoted to crystalline materials such as Quartz, $LiNbO_3$, $KTiOPO_4$, and α-BaB_2O_4 [19]. Nevertheless, recently the development of special glass compositions exhibiting NLO properties have extended the research into practical applications of glass transparent materials for a wide range of effects, such as fast intensity-dependent index, third-harmonic generation (THG), stimulated emission (or stimulated Raman scattering), second harmonic generation (SHG) and the multiphoton absorption [29]. Nonlinear phenomena in glasses, such as nonlinear refractive index, multiphoton absorption, and Raman and Brillouin scattering, depend on the glass itself, its nature (composition and structure), which is responsible for the nonlinearity. On the other hand, in glasses doped with RE ions or semiconductor nanoparticles, in which the glass assumes the role of host, the nonlinearity is produced by interactions between dopant ions, domains, and different phases (such as in glass-ceramics).

The first nonlinear effect in history is often associated with the beginning of the NLO [33], had occurred in 1875, when J. Kerr observed changes in the refractive index of a liquid (CS_2) in the presence of an electric field. The Kerr effect or quadratic electro-optic effect is directly related to the third-order nonlinearity, $\chi^{(3)}$. Pockels, 20 years later, observed another phenomena, the linear electro-optic effect [34], through the modification of the index of refraction of light in a

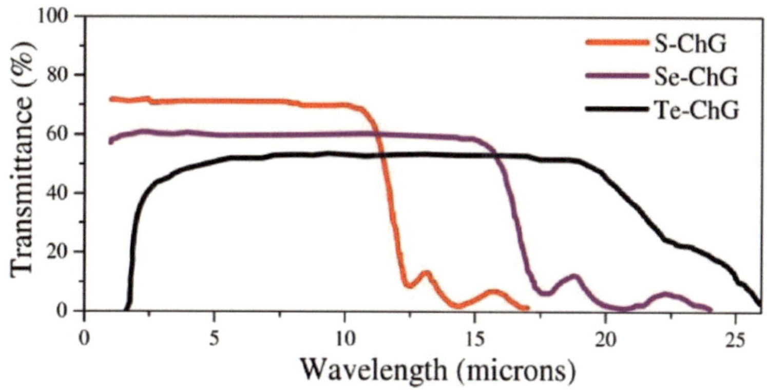

Figure 8.
Typical infrared (IR) transmission spectra of S-, Se-, and Te-based chalcogenide (ChG) glass [32].

non-centrosymmetric crystal (Quartz) placed by an electric field. For a long time thereafter, these phenomena were little studied and found of non-practical applications. However, the decisive prerequisite for work out such effects demands high laser pump intensities and suitable phase-matching conditions. Significant effects of NLO (e.g., frequency conversion by taking advantage of second and third harmonic generation) only began to be observed experimentally in the early 60s, after laser invention, due to the fact that such NLO effects require high electromagnetic field intensities to manifest, which was only possible using high-power lasers. P. Franken reported the first observation of the SHG in 1961 after focusing a pulsed ruby laser (λ = 694 nm) into a Quartz crystal; the red incident beam generated an emitted blue light (λ = 347 nm) [35]. THG was soon experimentally reported in 1965 [36]. Since the late of the 80s the interest in NLO properties in glass began to increase [19]. As already mentioned, the nonlinear optical response of glasses is closely related to their anionic polarizability [29, 37] which is described as the deformation of electron clouds (dipoles) when the electromagnetic field is applied. The selection of suitable glass structure and composition can contribute to efficiently optical Kerr effect, self-focusing, intensity-dependent refractive index, and other $\chi^{(3)}$ -related effects. In the literature, several reports have shown that the Kerr effect of non-conventional glass compositions is a viable option for self-phase modulation and broadband light generation in the near-infrared [29]. The $\chi^{(3)}$ in resonant mode is an additional possibility. Due to the bandwidth requirements for transmitting information for both long-haul and local area networks, Raman amplification is considered a good option to face out the recent developments in the telecommunications fiber industry and diode laser technology. Compared, for instance, with Er^{3+}-doped silica fiber amplifiers, in which the wavelength is fixed at 1550 nm, Raman gain bandwidths are larger, and the operational range only varies with the pump wavelength and the bandwidth of the Raman active medium (the glass nature) [29]. It is well known that the Kerr effect and Raman gain follow the polarity of the glass medium and are deeply impacted by the structure of some specific glasses, such as TeO_2 glass, which have large electronic polarizability. Additionality the small length of Te–O bond (2.01 Å) [37, 38] is considered responsible for the large third-order nonlinear optical susceptibility of these kinds of glass [38]. It $\chi^{(3)}$ value was as high as 1.4×10^{-12} esu about 50 times as large as that of SiO_2 glass [38].

The field of nonlinear optics of glasses has been mainly focused on two main groups: resonant and non-resonant [28]. Non-resonant interactions occur when the light excitation falls in the transparent wavelengths range of the glass longer than its electronic absorption edge. As no electronic transitions take place, the process can be seen as lossless and so an ultrafast glass response due to third-order electronic polarization is assured. Examples are, in general, high refractive index and high dispersion glasses like heavy flint optical glasses, or heavy metal oxide glasses, or chalcogenide glasses.

The resonant ones include semiconductor (quantum dots), or metallic nanoparticles doped glasses [10, 28] and the interaction occurs when the optical field's frequencies are near the electronic absorption edge so that its high resonant nonlinearity can be exploited. However, the isotropic structure glass and its amorphous state have inversion symmetry and do not exhibit second-order nonlinearity, $\chi^{(2)}$, or Pockels effect which is necessary for applications such as electro-optic switching and modulation or wavelength conversion in photonic technology. Indeed, glass is a good example of optically isotropic material (as well cubic crystals) that does not exhibit (in principle) any behavior that arises from that condition (e.g. optical birefringence). However, this is not always the case because second-order nonlinearity can be achieved in glass upon appropriate modification. For example, the application of both heat and electric fields (thermal poling) gives rise to SHG. Since $\chi^{(2)}$ is not physically possible in a centrosymmetric material, the creation of an axial symmetry under thermal poling has been demonstrated to be effective to introduce second-order nonlinearity properties [29]. Another route to create an optical SHG is by the introduction of optical non-linear nanocrystals within a glass matrix. Although thermal poling is an efficient way to induce SHG in silicate glasses, $\chi^{(2)}$ also appeared after glass heat treatments to precipitate crystallites of non-centrosymmetric compounds [39]. This strategy gives rise to transparent crystallized glasses (glass-ceramics). Nevertheless, more research is necessary to clarify some aspects, for instance, whether the thermal poling approach is effectively the best choice for raising SHG.

In the glass transparency region, which is found between the ionic (vibrational) and the electronic excitation interactions and where no permanent electric dipoles are present, the light frequency is too high for the ionic polarizability to follow the E field oscillations and too low to resonate with the electronic excitations [10]. Still, multiphoton processes may occur. For example, the probability of two-photon absorption is proportional to the square of the E field intensity [10].

4. Quantum dots doped glasses

Intensity-dependent nonlinear optical effects, such as the optical Kerr one, are very significant for all-optical data processing. Glasses with large nonlinear refractive index and nonlinear absorption coefficient are suitable materials for fiber telecommunication and nonlinear optical devices such as ultrafast optical switches and several photonic applications. Since silica and silicate glasses exhibit a small third-order nonlinear susceptibility $\chi^{(3)}$, the strategy of combining different materials to obtain composite systems, such as glass doped with semiconductor nanocrystals (quantum dots), allowed to obtain optimized nonlinear optical properties because semiconductors exhibit larger susceptibility. Glasses doped with semiconductors nanocrystals (quantum dots, QDs) such as CdS, CdSe, CdTe, PbS, CuCl, etc., are suitable materials for resonant NLO devices with response times on the ps domain. They can be prepared through the dispersion of a nanocrystalline phase in a glass matrix. This approach, through the reduction of bulk size to nanometric scale or quasi-zero-dimensional quantum dots, allow to change the electronic properties of glasses accordingly with enhanced nonlinearity compared with the corresponding bulk semiconductors [40]. Whenever the absorption of a photon of enough energy ($h\nu$ is greater than the band gap, E_g) excites an electron from the valence band to the conduction band in semiconducting materials, a free electron–hole pair may be formed. The hole and electron are attracted by Coulombic forces to keep them in a stable orbit as a bound electron–hole pair, called exciton [10]. Due to electrons and holes being confined in a small volume of radius, the radius of the exciton (distance between the electron and hole in an exciton), will change the available energy levels and the interaction with the photons. As the size of nanoparticles becomes progressively smaller, the quantum size effects of excitons confined in all three dimensions give rise to a series of discrete energy levels [10], and therefore the energy associated with them will depend on the relationship between the crystal size (R) and the exciton Bohr radius. Quantum confinement effects are quite significant in the range of $a \ll R \sim a_B$, where a is the lattice constant of the semiconductors, i.e. when R is similar to Bohr radius of exciton in bulk crystal (a_B). In QDs doped glasses these effects give rise to the so-called blue shift of the linear optical absorption edge. The shift regarding to the bulk E_g varies with R as $\sim 1/R^2$. Smaller R gives rise to larger blue-shift.

The size of semiconductor particles can be calculated by [41]:

$$\Delta E_g = h^2/8R^2\left[1/m_e^* + 1/m_h^*\right]$$
$$-1.8e^2/4\pi\varepsilon_0\varepsilon_\alpha R - 0.124e^4/\hbar(4\pi\varepsilon_0\varepsilon_\alpha)^2\left[1/m_e^* + 1/m_h^*\right]^{-1} \quad (17)$$

where ΔE_g is the shift of the band gap energy (due to the confinement), R is the particle size (radius), m_e^* and m_h^* are respectively the reduced effective masses of the electron (e) and hole (h). It is interesting to note that the second term, related to the kinetic energy of the electron and hole [41] exhibits a $1/R^2$ dependence while the third term, the Coulomb interaction between the electron and hole, has a 1/R dependence. Although the kinetic energy of the exciton for nanoparticles of $R \sim a_B$ seems to be predominant, the Coulomb interaction must also be considered [42]. **Figure 9** shown that the shift of the exciton resonances to higher energy (blue shift) is a consequence of the increasing quantum confinement as R decreases [43].

The changes in absorption also lead to refractive index changes, through the Kramers-Kronig transformation:

$$\Delta n(\omega) = \frac{c}{\pi}\int_0^\infty \frac{\Delta\alpha(\omega')}{\omega'^2 - \omega^2}d\omega' \quad (18)$$

where c is the speed of light and ω is the light frequency.

The method allows to correlate the determined change $\Delta\alpha$ in the absorption coefficient to the change Δn in the refractive index [43]. The nonlinear refractive index is then obtained by $n_2 = \Delta n/I$ (Eq. (8)). The value of $\chi^{(3)}$ will be proportional to the reciprocal of the confinement volume and will increase with decreasing R [10]. Is then expected that larger non-linearities are obtained for glasses containing smaller particles and larger volume fractions of QDs [10].

5. Metal-doped glasses

Metal doped glass possesses linear and nonlinear optical properties. Great interest has driven the study of the third-order nonlinear susceptibility of metal particles embedded in dielectric matrices, like glasses [44], which are influenced not only by the type and size of the metal particles but also by the metal-dielectric constant. The most significant effect of the confinement of metal particles in optical properties of nanocomposite glasses is the appearance of the surface plasmon resonance, which

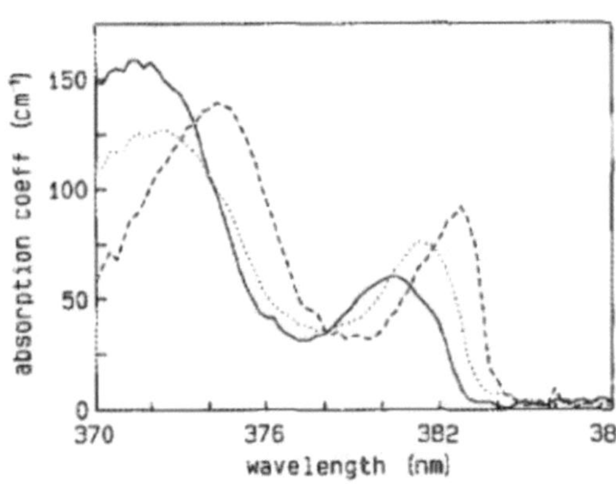

Figure 9.
Absorption spectra of CuCl-doped quantum dot glasses: 22 Å (solid); 27 Å (dot); 34 Å (dash) [43].

deeply enhances the glass $\chi^{(3)}$ responses with picosecond temporal responses. For example, the optical absorption spectrum of Ag-doped silica sol–gel glass shows the presence of an absorption band of surface plasmon resonance due to Ag nanoparticles at ~420 nm (**Figure 10**).

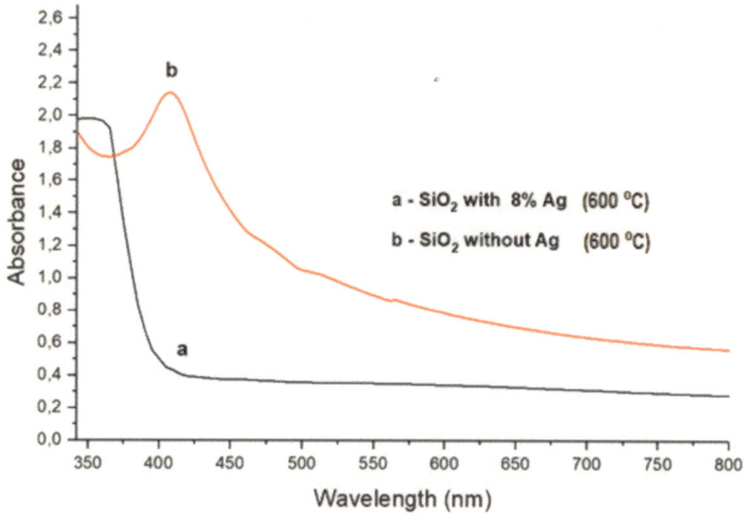

Figure 10.
Absorption spectra of Ag-SiO2 cermet (at a concentration of 8% Ag) and SiO2 matrix (without Ag).

Plasmons deals with a coherent interaction between the free-electron gas surrounding metal and the incident radiation. The motion of these free electrons can be described by the plasma Drude model, along with a plasma frequency of the bulk metal ω_P. In accordance with the Drude free-electron model, the dielectric constant of metal particles is given by [45]:

$$\varepsilon_m = \varepsilon'_m - i\varepsilon''_m = 1 - \omega_P^2/[\omega(\omega - i/\tau)] \tag{19}$$

Where τ is the time between collisions among electrons. The real (ε’) and imaginary (ε”) parts of the complex dielectric constant are expressed as [45]:

$$\varepsilon'_m = n^2 - k^2 = 1 - \omega_P^2\tau^2/\left[1 + (\omega\tau)^2\right] \tag{20}$$

$$\varepsilon''_m = 2nk = \omega_P^2\tau/\left\{\omega\left[1 + (\omega\tau)^2\right]\right\} \tag{21}$$

From the above equations is possible to infer the existence of an interaction between the free-electron gas and the incident electromagnetic field, which gives rise to an excitation of the electrons at the metal surface, associated with collective oscillations of electrons in the metal nanoparticles, called surface plasmon. The large value of $\chi^{(3)}$ of metal-doped glasses arises predominantly from the local electric field enhancement near the surface of the metal nanoparticles (Ag, Cu, Ni, or other metal nanoparticles) due to their surface plasma resonance, leading to a variety of optical effects.

When the diameter (d) of metal particles is much lower than the wavelength of light (λ), scattering is negligible. As well, the total collisional impacts of the electrons with the particle surfaces become significant and a new-found relaxation time, τ_{eff}, appeared, given by [45]:

$$1/\tau_{eff} = 1/\tau_b + 2v_F/d \tag{22}$$

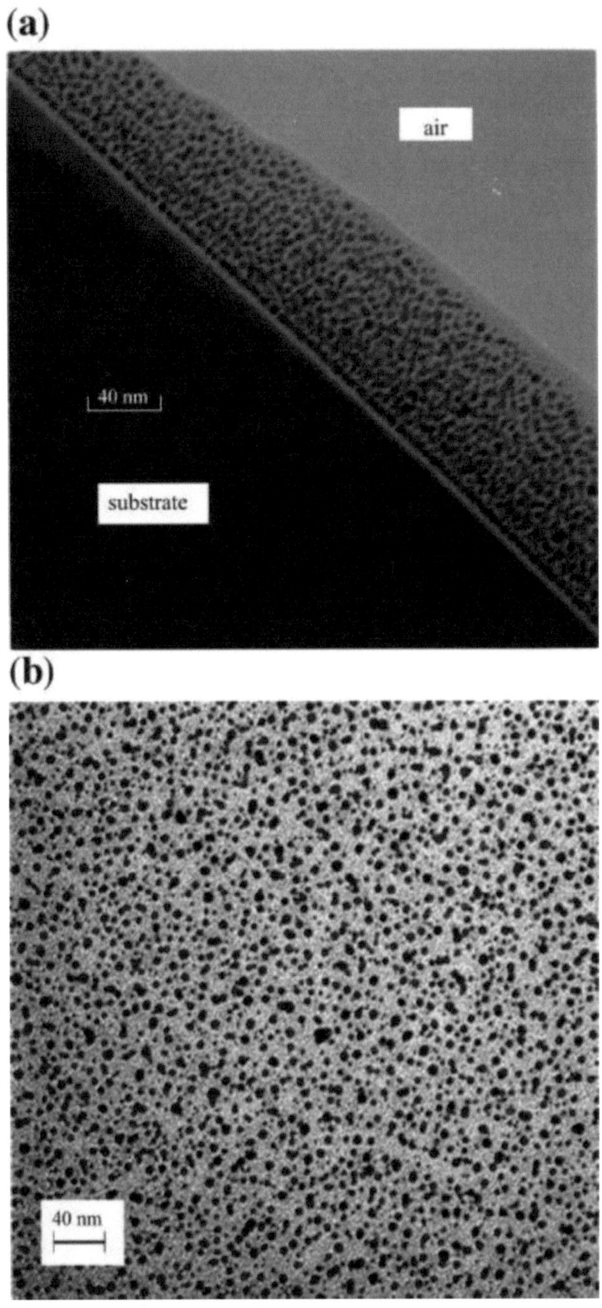

Figure 11.
Transmission electron microscopy micrographs of Au-SiO2 thin films: a) cross section view of a film with Au volume fraction p = 23%, and b) plan view of a film of Au volume fraction p = 8% [46].

where τ_b is the bulk value and v_F is the electron velocity at the Fermi energy. Spherical metal nanoparticles embedded in a glass matrix with a real dielectric constant ε_d exhibit NLO properties. **Figure 11** exhibits homogeneous size distribution of spherical Au nanoparticles in a SiO_2 thin film on a metal substrate [46]. For the conditions.

The equation usually considered to obtain the $\chi^{(3)}$ of metal/glass composites, is given by [45]:

$$\chi^{(3)} = 3pf^4\chi_m^{(3)} \tag{23}$$

Where $\chi_m^{(3)}$ is the bulk metal third-order susceptibility, f is the local electric field near the metal particles and p is the metal volume fraction. The optical response of metal particle/glass composites can be determined by the local field enhancement inside the nanoparticles (dielectric confinement):

$$f = \frac{E}{E_0} = \frac{3\varepsilon_d}{\varepsilon_m + 2\varepsilon_d} \tag{24}$$

f is given by the ratio between the field E inside a metal particle and the applied field E_0, with ε_d the dielectric constant of the glass matrix and ε_m the one of the metal.

So, if one assumes $\chi_m^{(3)}$ independent of particle size, then $\chi^{(3)}$ will increase as the volume fraction of metal particles and their size increases [45].

6. Conclusions

In the last decades, the development of optics, as the science that deals with light and its applications, has had an enormous growth not only through new or recognized theoretical concepts but also in new optical techniques and new instruments. Several factors contributed to this, namely: 1) the emergence of new light sources, such as lasers, which allowed the advent of new applications associated with light manipulation, such as those based on the nonlinear optical properties of materials; and 2) the development of new glasses or the modification/optimization of others through the addition of dopants (e.g., metallic nanoparticles or QDs), also allowed the creation of new photonic devices (light sources, all-optical switches, modulators, etc.) and new technologies associated with them. These developments also gave rise to the so-called integrated optics, which allowed a reduction in the size of optical systems, while maintaining their high nonlinear optical performance. Many of these technologies are used in the field of communications and other sectors of activity, such as health and information. In terms of materials, NLO glasses have grown as indicated by the numerous scientific publications on the subject. Glasses have great versatility and offer great flexibility to modify their nonlinear responses by manipulating their composition, refractive index, gap, etc. Because of their structural inversion symmetry, glasses do not possess second-order optical nonlinearity. Yet, it is possible to induce this optical response in the glass by thermal electric poling.

Conflict of interest

The author declares no conflict of interest.

Author details

Helena Cristina Vasconcelos[1,2]

1 Faculty of Sciences and Technology, Azores University, Ponta Delgada, Portugal

2 Laboratory of Instrumentation, Biomedical Engineering and Radiation Physics (LIBPhys-UNL), Department of Physics, NOVA School of Science and Technology, NOVA University Lisbon, Caparica, Portugal

*Address all correspondence to: helena.cs.vasconcelos@uac.pt

References

[1] Poole SB, Payne DN, Fermann ME. Fabrication of low-loss optical fibres containing rare-earth ions. Electronics Letters. The Institution of Engineering and Technology. 1985;**21**:737-738. DOI: 10.1049/el:19850520

[2] Refi JJ. Optical fibers for optical networking. Bell Labs Technical Journal. 1999;**4**(1):246-261. DOI: 10.1002/bltj.2156

[3] Hughes M. Modified Chalcogenide Glasses for Optical Device Applications. 2007

[4] Vasconcelos HC, Pinto AS. Fluorescence properties of rare-earth-doped sol-gel glasses. In: Chandra U, editor. Recent Applications in Sol-Gel Synthesis. Rijeka: IntechOpen; 2017. DOI: 10.5772/intechopen.68534. Available from: https://www.intechopen.com/chapters/55130

[5] Islam MN. Raman amplifiers for telecommunications. IEEE Journal of Selected Topics in Quantum Electronics. 2002;**8**(3):548-559. DOI: 10.1109/JSTQE.2002.1016358

[6] Toulouse J. Optical nonlinearities in fibers: Review, recent examples, and systems applications. Journal of Lightwave Technology. 2005;**23**(11):3625-3641. DOI: 10.1109/JLT.2005.855877

[7] Available from: http://www.physics.ttk.pte.hu/files/TAMOP/FJ_Nonlinear_Optics/10_nonlinear_fiber_optics.html

[8] Miller SE. Integrated optics: An introduction. The Bell System Technical Journal. 1969;**48**(7):2059-2069. DOI: 10.1002/j.1538-7305.1969.tb01165.x

[9] Sirleto L, Ferrara MA. Fiber amplifiers and fiber lasers based on stimulated raman scattering: A review. Micromachines (Basel). 2020;**11**(3):247. DOI: 10.3390/mi11030247

[10] Available from: https://www.lehigh.edu/imi/teched/OPG/lecture37.pdf

[11] Available from: https://www.brown.edu/research/labs/mittleman/sites/brown.edu.research.labs.mittleman/files/uploads/lecture35_0.pdf

[12] Yariv A. Quantum Electronics. Third ed. New York: John Wiley & Sons; 1989

[13] Liu L. Second-Order Optical Nonlinear Properties of Glasses in Photonic Glasses. Singapore: World Scientific; 2006. pp. 153-189. DOI: 10.1142/ 54 9789812773487_0005

[14] Available from: https://www.rp-photonics.com/nonlinear_index.html

[15] Pelusi MD et al. Applications of highly-nonlinear chalcogenide glass devices tailored for high-speed all-optical signal processing. IEEE Journal of Selected Topics in Quantum Electronics. 2008;**14**(3):529-539. DOI: 10.1109/JSTQE.2008.918669

[16] Vasconcelos HC, Gonçalves MC, editors. Overall Aspects of Non-Traditional Glasses. Synthesis, Properties and Applications. Sharjah, U.A.E: Bentham Science Publishers. ISBN: 978-1-68108-208-0 Hardcover, eISBN: 978-1-68108-207-3; 2016

[17] Nasu H, Uchigaki T, Kamiya K, Kanbara H, Kubodera K. Nonresonant-type third-order nonlinearity of (PbO, Nb2O5)-TiO2-TeO2 glass measured by third-harmonic generation. Japanese Journal of Applied Physics. 1992;**31**:3899-3900

[18] Agrawal GP. Chapter 11 - Highly nonlinear fibers. In: Agrawal GP, editor. Nonlinear Fiber Optics. Sixth ed. Academic Press. NY, USA: University of Rochester. 2019. pp. 463-502. ISBN

9780128170427. DOI: 10.1016/B978-0-12-817042-7.00018-X

[19] Dussauze M, Cardinal T. Nonlinear optical properties of glass. In: Musgraves JD, Hu J, Calvez L, editors. Springer Handbook of Glass. Springer International Publishing; 2019. pp. 157-189. Springer Handbooks, 978-3-319-93726-7. ff10.1007/978-3-319-93728-1ff. ffhal-02309707. https://hal.archives-ouvertes.fr/hal-02309707

[20] Van Stryland EW, Sheik-Bahae M. Z-scan technique for nonlinear materials characterization. In: Materials Characterization and Optical Probe Techniques: A Critical Review. SPIE-International Society for Optics and Photonics. 1997:102910Q. Available from: https://ui.adsabs.harvard.edu/abs/1997SPIE10291E.0QV. DOI: 10.1117/12.279853

[21] Available from: https://www.lehigh.edu/imi/teched/OPG/lecture12.pdf

[22] Parker JM. Glasses. In: Bassani F, Liedl GL, Wyder P, editors. Encyclopedia of Condensed Matter Physics. Elsevier; 2005. pp. 273-280. ISBN 9780123694010. DOI: 10.1016/B0-12-369401-9/00538-6

[23] Cui S, Chahal R, Boussard-Plédel C, Nazabal V, Doualan J-L, Troles J, et al. From selenium- to tellurium-based glass optical fibers for infrared spectroscopies. Molecules. 2013;**18**(5):5373-5388. DOI: 10.3390/molecules18055373

[24] Karasu B, İdinak T, Erkol E, Yanar AO. Chalcogenide glasses. El-Cezeri. 2019;**6**(3):428-457. DOI: 10.31202/ecjse.547060

[25] Almeida JMP, Barbano EC, Arnold CB, Misoguti L, Mendonça CR. Nonlinear optical waveguides in As2S3-Ag2S chalcogenide glass thin films. Optical Materials Express. 2017;**7**: 93-99

[26] Chakraborty P. Metal nanoclusters in glasses as non-linear photonic materials. Journal of Materials Science. 1998;**33**(9):2235-2249. DOI: 10.1023/a:1004306501659

[27] Dimitrov V, Komatsu T. Classification of oxide glasses: A polarizability approach. Journal of Solid State Chemistry. 2005;**178**(3):831-846. DOI: 10.1016/j.jssc.2004.12.013

[28] Yamane M, Asahara Y. Nonlinear optical glass. In: Glasses for Photonics. Cambridge: Cambridge University Press; 2000. pp. 159-241. DOI: 10.1017/CBO9780511541308.005

[29] Cardinal T, Fargin E, Videau JJ, Petit Y, Guery G, Dussauze M, et al. Glass and glass ceramic for nonlinear optics: Fundamentals to applications. In: Functional Glasses: Properties And Applications for Energy and Information. H. Jain, Lehigh Univ.; C. Pantano, The Pennsylvania State Univ.; S. Ito, Tokyo Institute of Technology; K. Bange, Schott Glass (ret.); D. Morse, Corning Eds, ECI Symposium Series. 2013. Available from: https://dc.engconfintl.org/functional_glasses/14

[30] Krogstad MR. "Ge-Sb-Se Chalcogenide Glass for Near- and Mid-Infrared Nonlinear Photonics". Thesis (Ph.D., Physics). USA: University of Colorado; 2017. Available from: https://scholar.colorado.edu/concern/graduate_thesis_or_dissertations/4x51hj00j

[31] Ono M, Hata M, Tsunekawa M, et al. Ultrafast and energy-efficient all-optical switching with graphene-loaded deep-subwavelength plasmonic waveguides. Nature Photonics. 2020; **14**:37-43. DOI: 10.1038/s41566-019-0547-7

[32] Gao S, Bao X. Chalcogenide taper and its nonlinear effects and sensing applications. iScience. 2020;**23**(1): 100802. DOI: 10.1016/j.isci.2019.

100802. Epub 2019 Dec 25. PMID: 31927486; PMCID: PMC6957858

[33] Fowles G. Introduction to Modern Optics. 2nd ed. New York: Dover Publications; 1989

[34] Narasimhamurti TS. Photoelastic and Electro-Optic Properties of Crystals. New York: Plenum; 1981

[35] Franken PA, Hill AE, Peters CW, Weinreich G. Generation of optical harmonics. Physical Review Letters. 1961;**7**:118

[36] Maker PD, Terhune RW. Study of optical effects due to an induced polarization third order in the electric field strength. Physics Review. 1965;**137**: A801-A818

[37] Azlan MN, Halimah MK, Shafinas SZ, Daud WM. Electronic polarizability of zinc borotellurite glass system containing erbium nanoparticles. Materials Express. 2015;**5**(3):211-218. DOI: 10.1166/mex.2015.1236

[38] Kim S-H, Yoko T, Sakka S. Linear and nonlinear optical properties of TeO2 glass. Journal of the American Ceramic Society. 1993;**76**:2486-2490. DOI: 10.1111/j.1151-2916.1993.tb03970.x

[39] Guignard M, Nazabal V, Zhang X, Smektala F, Moréac A, et al. Crystalline phase responsible for the permanent second-harmonic generation in chalcogenide glass-ceramics. Optical Materials. 2007;**30**(2):338-345. DOI: ⟨10.1016/j.optmat.2006.07.021⟩. ⟨hal-00172320⟩

[40] Banfi GP, Degiorgio V, Ricard D. Nonlinear optical properties of semiconductor nanocrystals. Advances in Physics. 1998;**47**(3):447-510. DOI: 10.1080/000187398243537

[41] Prabhu RR, Khadar MA. Characterization of chemically synthesized CdS nanoparticles. Pramana-Journal of Physics. 2005;**65**: 801-807. DOI: 10.1007/BF02704078

[42] Lippens PE, Lannoo M. Calculation of the band gap for small CdS and ZnS crystallites. Physical Review B: Condensed Matter. 1989;**39**(15): 10935-10942. DOI: 10.1103/ physrevb.39.10935

[43] Justus BL, Seaver ME, Ruller JA, Campillo AJ. Applied Physics Letters. 1990;**57**:1381-1383

[44] Kim K-H, Husakou A, Herrmann J. Linear and nonlinear optical characteristics of composites containing metal nanoparticles with different sizes and shapes. Optics Express. 2010;**18**: 7488-7496

[45] Available from: https://www.lehigh.edu/imi/teched/OPG/lecture39.pdf

[46] Pinçon-Roetzinger N, Prot D, Palpant B, Charron E, Debrus S. Large optical Kerr effect in matrix-embedded metal nanoparticles. Materials Science and Engineering: C. 2002;**19**(1–2):51-54. DOI: 10.1016/S0928-4931(01)00431-3

Section 4

Novel Application of Nonlinear Schrodinger Equation

Chapter 6

Deep Saturation Nonlinearity of 5G Media and Potential Link to Covid-19

Mohsen Lutephy

Abstract

5G broadband millimeter LFs (low frequencies) are filtered and do not influence into the cells, but in the nonlinear media, the modulation instability of the fast underlying carrier wave leads to appear the slowly varying perturbation parasite envelopes (noises) which is described by nonlinear Schrodinger equation (NLSE). Thus, the 5G pump waves in nonlinearity leads to extremely low frequency electromagnetic pulse envelopes enable to pass the filters such as the skin, and disintegrating in the cells to the 5G carrier waves and disordering genome as a probable origin to organize the corona virus via covering separated part of the genome with the capsids. A so called physical solution on the modulation instability of the nonlinear media is the Kuznetsov-Ma breather revealed previously in the optical fibers and accordingly we have detected here the signature of the Kuznetsov-Ma breather self-similar solution of the NLSE on the global distribution pattern of the covid-19 infection and death cases as an agreement between the theoretical results and observations for covid-19. A possible potential link between the covid-19 and 5G nonlinear internet media is revealed, verifying that the covid-19 global patterns of the infection and death cases are statistically significant.

Keywords: Nonlinear optics, Radiation health risk, Epidemiology, LF radiobiology, covid-19, 5G

1. Introduction

5G internet media uses millimeter waves (10–300 GHZ) in which are easily blocked environmentally and not travel far. Despite previous generations of wireless media, the 5G multiple antennas arrange in "phased arrays" [1, 2] that work together to emit focused, steerable, laser-like beams that track each other causes to appear nonlinear dispersive media, environmentally revealing modulation instabilities due to nonlinearity which is observable in the optical fibers e.g. [3–5] (there are many papers to cite here).

In the linearity, the plane wave is ever stable and perturbations do not grow whereas that in the nonlinearity, the perturbation in the background plane wave is expanded and can produce high amplitude perturbation parasites (noises) via the modulation instabilities which yields to its relevant wave solutions.

Bill P. Curry in his so-called graph reported exponential Microwave absorption in brain tissue, newly confirmed more in the paper "Exposure of Insects to Radio-Frequency Electromagnetic Fields from 2 to 120 GHz" [6]. However the 5G waves

IntechOpen

are affecting the cells in the long-term exposure e.g. [7–14] (the articles published in the field of the cells at the exposure of the LF EMF is out of capacity all to cite here) and also bioeffects at exposure of modulated EMF (MEMF) of radiofrequencies of low intensity e.g. [15] and reference in, it is yet correct that the 5G waves are blocked by skin and rarely influence in the bodies is for radio entropy. Then it has been thought that "The 5G Health Hazard That Isn't" as titled in the New York Times. It is real also the radio waves become safer at higher frequencies for that easier to block environmentally and also by skin. But according to the reports, the cells exposed to extremely low frequencies (ELFs) presented an increase of the number of cells with high damaged DNA as compared with non-exposed cells. Then if the 5G media was linear, the 5G electromagnetic waves were almost safe for healthiness of the species in the earth, but since the 5G media is nonlinear then for modulation instability of the nonlinear media, the 5G wireless internet produces extremely low frequency perturbation parasite pulses similar to the perturbation pulses produced in the fiber optics.

Some properties of nonlinearity are appeared in nonlinear media such as self-focusing [16, 17] and wave steepening [18]. The nonlinear dispersive media is described by the nonlinear Schrodinger equation (NLSE) discussed in many papers and the books e.g. [19] as observed in the deep-water wave propagation [20, 21] and optical fiber [22, 23].

NLSE is a central model of nonlinear science, applying to hydrodynamics, plasma physics, molecular biology and optics. NLSE describes the slowly varying envelope that modulates a fast underlying carrier wave. The perturbation method on the NLSE yields to the wave-packet envelopes as the parasite for 5G low frequency carrier waves, enable to cross the skin and expose the cells in the field of extremely low frequencies (ELFs). Thus, the 5G background pump waves can inter noticeable to the cells and damaging DNA and producing some disorders may be relevant to the corona virus which is a part of genome covered by the capsids (capsidal DNA). If this is true it should be an identity between the distribution patterns of the corona virus infection and death cases and wave pattern of the 5G perturbation parasite envelopes.

Wonderfully we detect here the identity between the covid-19 distribution pattern of the infection and death cases and the Kuznetsov-Ma breather self-similar solution of the nonlinear Schrodinger equation (NLSE) which describes the instability of the radio frequency waves in dispersive nonlinear media.

2. 5G and nonlinear Schrodinger equation and covid-19

2.1 Theoretical frameworks

The envelope evolution of nonlinear systems is extracted in the form of nonlinear Schrodinger equation (NLSE) from different methods such as Fourier-Mode coupling e.g. [24], or Multiple-Scale analysis [25]. The NLS equation in nonlinear optics was first derived by Kelley in 1965 using a nonlinear electromagnetic wave Maxwell's equation introduced by Chiao et al. one year earlier [26, 27]. Furthermore, Karpman and Krushkal in 1969 derived the NLS equation using Whitham-Lighthill adiabatic approximation [28].

Consider a system described by the normalized nonlinear Schrodinger equation [$(1 + 1)$D NLSE]:

$$i\frac{\partial\varphi}{\partial t}+\frac{1}{2}\frac{\partial^2\varphi}{\partial x^2}+f\left(|\varphi|^2\right)\varphi=0 \tag{1}$$

Eq. (1) describes many physical systems, primarily those in which nonlinear waves propagate in isotropic media [29]. The parameter φ describes the slowly varying envelope which modulates a fast underlying carrier wave and the nonlinear term $f\left(|\varphi|^2\right)$ is specific to the physical system.

On the (1 + 1)D NLSE, two particular forms of nonlinearity are Kerr-type and saturable type that are common in optics [30]. The Kerr-type, where $f\left(|\varphi|^2\right) = |\varphi|^2$ and the saturable type, where $f\left(|\varphi|^2\right) = |\varphi|^2/\left(1+|\varphi|^2\right)$.

Of course where $|\varphi|^2/\left(1+|\varphi|^2\right) \approx 1 - \left(1/|\varphi|^2\right)$, the nonlinearity is called the "deep saturation nonlinearity" [31] and most of the interesting properties and energy of the solitons is in the regions where $|\varphi|^2 > > 1$ [31].

If $\varphi(\xi, \tau)$ is a solution of the (1 + 1)D NLSE in deep saturation nonlinearity, then a whole family of the solutions is obtained [31] by re-scaling via real parameter ε as

$$\varphi(\xi, \tau) \rightarrow e^{it\left(1-\varepsilon^2\right)}\varepsilon^{-1}\varphi\left(\varepsilon\xi, \varepsilon^2\tau\right) \tag{2}$$

Then all solutions of the same order of (1 + 1)D deep saturable NLSE are related each other by rescaling and the solutions are self-similar to one another in their physical properties, such as intensity, shape, etc. "This is because the natural scale in the saturable NLSE is visible only in the margins of the intensity profile of the soliton, and its effect on the shape is tiny" [31].

On the Nonlinear Schrodinger equation, the modulation instability of the background plane wave motivates to create localized breathers [32]. Of course there are various types of the numerical and analytical solutions describing breathers relevant to the definition of the nonlinear term $f\left(|\varphi|^2\right)$, but qualitatively breathers can be modeled in a specific solution [33, 34] as the localized pulses on the continuum background plane wave that:

$$\varphi = \left[\frac{2(1-2a)\cosh\left(\sqrt{8a(1-2a)}\tau\right) + i\sqrt{8a(1-2a)}\sinh\left(\sqrt{8a(1-2a)}\tau\right)}{\sqrt{2a}\cos\left(2\sqrt{1-2a}\xi\right) - \cosh\left(\sqrt{8a(1-2a)}\tau\right)}\right] \tag{3}$$

Where a < 1/2, the solution is Akhmediev breather and for a = 1/2 the solution is peregrine soliton and for a > 1/2 it is Kuznetsov-Ma breather. Of course amplitude of the background wave has been deleted here for reality that the background plane wave is without modulation and thus, the envelopes do work as the independent wave packets.

First solution of the NLSE was the Kuznetsov-Ma (KM) breather [35, 36] and we consider also a Kuznetsov-Ma breather type solution for perturbation of the carrier wave in deep saturation nonlinearity. This solution does show the localized pulsation of the background wave as the periodic noise.

The experimental results in the optical fibers e.g. [5] verify that the Kuznetsov-Ma breather is matched with real term $\mathrm{Re}\{\varphi\}$ and in fact the imaginary term $\mathrm{Im}\{\varphi\}$ can be imaginary and then the physical answer is real term of the Eq. (3), that is,

$$\varphi = \left[\frac{2(1-2a)\cos\left(\sqrt{8a(2a-1)}\tau\right)}{\sqrt{2a}\cos\left(2\sqrt{1-2a}\xi\right) - \cos\left(\sqrt{8a(2a-1)}\tau\right)}\right] \tag{4}$$

Notice that $\cosh(i\tau) = \cos(\tau)$ and then in the Kuznetsov-Ma breather solution in which a > 1/2, we have $\cosh\left(\sqrt{8a(1-2a)}\tau\right) = \cos\left(\sqrt{8a(2a-1)}\tau\right)$.

However if we use the φ included to both of the imaginary and real terms, the solution numerically is still near to the Eq. (4) and difference is neglect-able.

2.2 The evidences for the theoretical arguments

5G sideband waves are background carrier wave which in the nonlinearity yields to the relevant instabilities. The description of instabilities in optics as rogue waves is recent, however, first used in 2007 when shot-to-shot measurements of fiber supercontinuum (SC) spectra by Solli *et al.* yielded long-tailed histograms for intensity fluctuations at long wavelengths [37].

On the modulation instability of the 5G media, however the 5G monochromatic waves do not directly affect the cells but producing extremely low frequency envelope impulses which are enable to influence into the cells. The scale invariance in the deep saturation nonlinearity according to the Eq. (2) yields to the answer as the full family of the self-similar solutions (intensity in the term $|\varphi|^2$) that

$$|\varphi|^2 = \sum_{n=0}^{+\infty} \left|\varepsilon^{-n}\varphi\left(\varepsilon^n \xi, \varepsilon^{2n}\tau\right)\right|^2 \tag{5}$$

Very short period waves are neglected actually and very long waves are trivial in short time (in the scale of lesser than several years) and then actually we can consider a solution as a suit of four consequent envelopes from the Eq. (5) in series of $n = \{0, 1, 2, 3\}$.

The time τ is free scale mathematically and we can rescale $\tau + \pi/2 \to \tau\sqrt{8a(2a-1)}$. We assume arbitrary the frame center at point $\xi = 0$ and then by the Eqs. (4), (5) we deduce for $n = \{0, 1, 2, 3\}$, an actual answer as follows

$$|\varphi|^2 = \left[\varepsilon^{-3}\left|\frac{2(2a-1)\sin\left(\varepsilon^6\tau\right)}{\sqrt{2a} - \sin\left(\varepsilon^6\tau\right)}\right| + \varepsilon^{-2}\left|\frac{2(2a-1)\sin\left(\varepsilon^4\tau\right)}{\sqrt{2a} - \sin\left(\varepsilon^4\tau\right)}\right| + \varepsilon^{-1}\left|\frac{2(2a-1)\sin\left(\varepsilon^2\tau\right)}{\sqrt{2a} - \sin\left(\varepsilon^2\tau\right)}\right| + \left|\frac{2(2a-1)\sin(\tau)}{\sqrt{2a} - \sin(\tau)}\right|\right]^2 \tag{6}$$

Figure 1.
Theoretical wave (self-similar corona-like) solution of covid-19 infection pattern for $a = 1, \varepsilon = 1/4, k_I = 1/5$ driven by the Eqs. (6), (7).

If corona virus infection cases per time symbolized here by n_I relates to 5G wave's breather amplitude $|\varphi|^2$, thus there should be exist a coefficient k_I to correlate the intensity of 5G perturbation envelopes to the infection of the corona virus as

$$n_I = k_I|\varphi|^2 \tag{7}$$

The infection cases per time interval symbolized here with n_I in unit K (cases/time) is drawn on the Eqs. (6), (7) in **Figure 1** with the best fit in $\varepsilon = 1/4$, a = 1, k_I = 1/5 (K in vertical axis and time in horizontal axis in the radians scale). Transferring the time from radians to the date, the wave pattern matches nicely with covid-19 infection pattern reported by live such as www.worldometers.info. The domain and time scales of the wave envelopes match with covid-19 infection pattern as compared in the **Figure 2**.

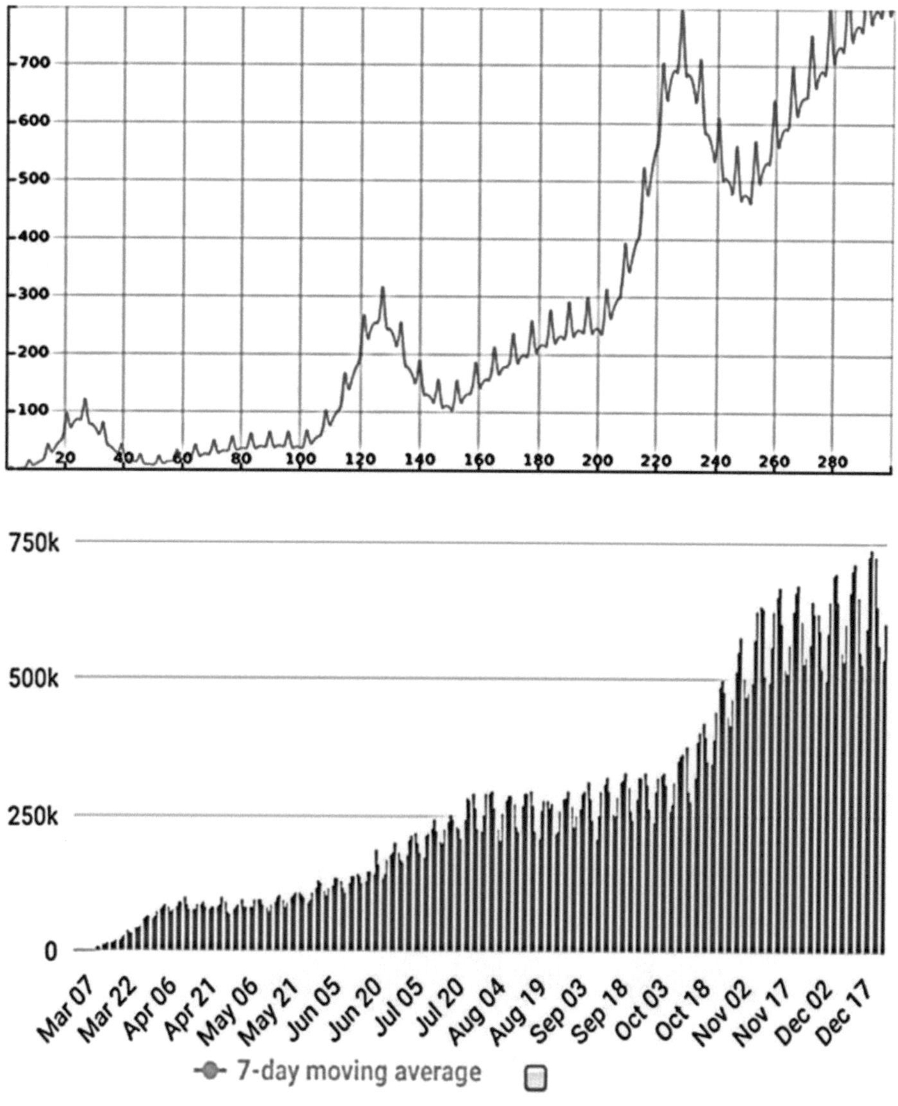

Figure 2.
Covid-19 infection pattern to date (below diagram) compared to 5G perturbation pulses in the time scaled with radians (above diagram).

The corona virus is transmitted by human to human biologically and thus, the biologic spreader effect moderates extremes of the physical waves. For example if the antennas do not work for a short time, still the covid-19 is continued for moderation of the biological effects. Of course deviation from the physical source will appear along the time. In reality the 5G internet media includes the antennas distributed in the earth and thus, the phase of the perturbation envelopes can vary by changes in the antennas.

By the way via comparing the covid-19 infection diagram with the wave solution Eq. (6) along the year 2020, still the covid-19 is matched with the Kuznetsov-Ma wave solution of the NLSE both in the phase and shape of the wave. We have drawn the pure wave solution (above diagram in the **Figure 2.**) without moderation of the extremes and then difference in the intensity between the diagrams of the wave solution and covid-19 infection pattern is natural. In reality the above diagram in the **Figure 2** which is for Kuznetsov-Ma breather should be moderated to fit in the intensity with below diagram in the **Figure 2** which is for covid-19 infection cases. The 5G injection effect of the parasite pulses in the cells does not depend to the cases-age but deaths per time interval symbolized here by n_D depends to the

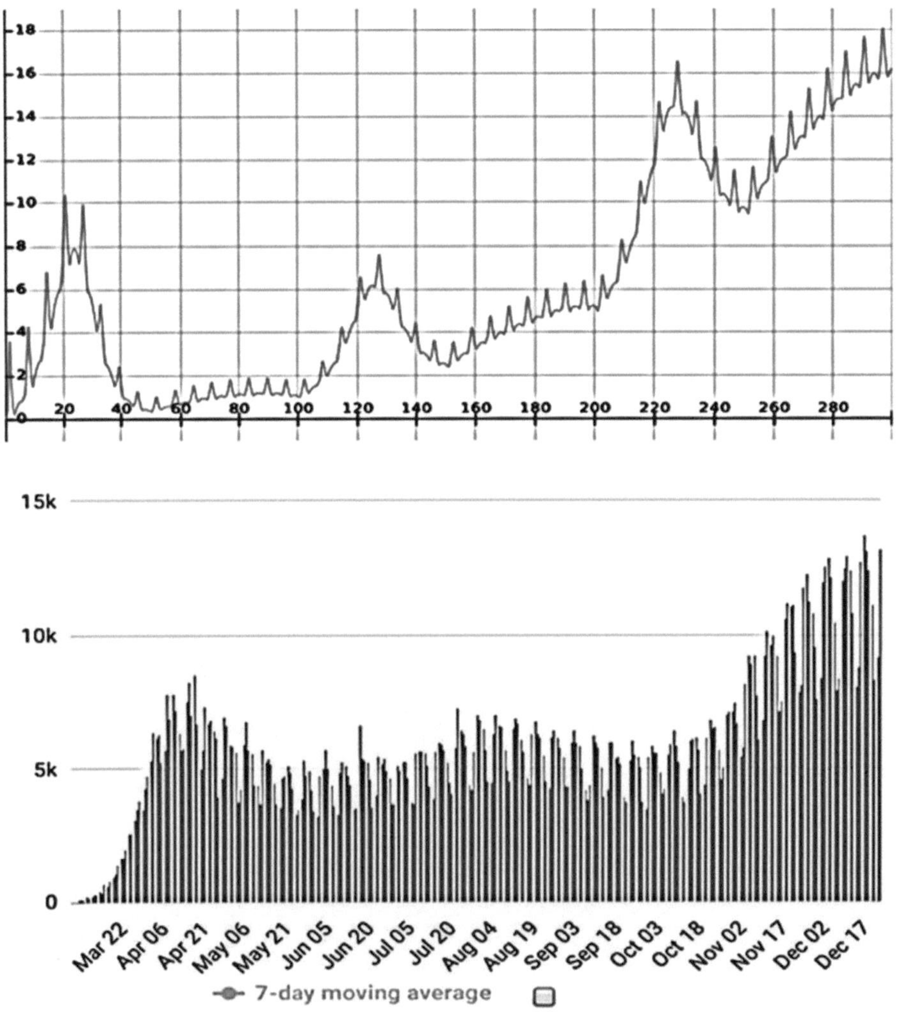

Figure 3.
Covid-19 death cases pattern (below diagram which is copied from online corona-meter web) compared with the ϕ^2 *from Eq. (6).*

cases-age. The index n_D/n_I is larger for elders. The population of elder cases is decreased proportionally along the time and then the index n_D/n_I is flattening rapidly to asymptotic size ~ 0.02. We find numerically a function for damping effect of the index n_D/n_I as follows

$$n_D/n_I = \left[0.02 + \frac{‘1}{1 + 0.3\tau + 0.005\tau^2}\right] \tag{8}$$

An initial shock wave is observed for death cases in onset of the covid-19 pandemic which is flattening asymptotically to the normal size. This is matched with daily deaths observed in covid-19 (**Figure 3**).

The wireless antennas have potential to transfer the earth to the nonlinear media as the source of extremely low frequency (ELF) electromagnetic envelopes (Trojan horse) affecting the cells, disordering the genome and damaging the species which may be visible in the next generations the more.

3. Conclusions

However the 5G broadband electromagnetic pump waves are blocked mainly and environmentally disable to pass the skin but on the nonlinearity of 5G internet media as a result of modulation instability we find extremely low frequency perturbation parasites as the solution of nonlinear Schrodinger equation (NLSE), enable to pass the skin and disordering the genome as the potential link between the 5G and covid-19. Accordingly the global patterns of the corona virus infection and death cases follow the Kuznetsov-Ma type breather solution of the NLSE which is a perturbation envelope. In detail we find that the global patterns (infection and death) of the corona virus is matched with deep saturation nonlinearity of 5G internet broadband media which yields to a self-similarity of the infection and death cases global patterns of the corona virus.

Acknowledgements

We need to thanks from human right organizations and scientists and doctors and generally people for their warnings for health risks by EMF which is increasing every day and have transferred the earth to an electromagnetic bathroom and nonlinear dangerous media which increases the health risks globally.

Conflict of interest

The authors declare no conflict of interest.

Author details

Mohsen Lutephy
Islamic Azad University (IAU), Tehran, Iran

*Address all correspondence to: lutephy@gmail.com

References

[1] Hong W, Jiang ZH, Yu C, et al. Mint: Multibeam antenna technologies for 5G wireless communications. IEEE Tr Ant Prop. 2017;**65**(12):6231-6249

[2] Chou H-T. Mint: Design Metodology for the Multi-Beam phased Array of antennas with Relatively Arbitrary Coverage Sector. Conference paper: 2017 11th European Conference on Antennas and Propagation; Paris, France. DOI: 10.23919/EuCAP.2017.7928095

[3] Dudley JM, Genty G, Dias F, Kibler B, Mint AN. Modulation instability, Akhmediev Breathers and continuous wave supercontinuum generation. Opt. Express. 2009;**17**: 21497-21508

[4] Van Simaeys G, Emplit Ph, & Haelterman M. Mint: Experimental demonstration of the Fermi-Pasta-Ulam recurrence in a modulationally unstable optical wave. Phys. Rev. Lett. 2001;87, 033902

[5] Kibler B, Fatome J, Finot C, Millot G, Dias F, Genty G, et al. Mint: The Peregrine soliton in nonlinear fibre optics. Nature Physics. 2010;**6**:795

[6] Thielens A, Bell D, Mortimore DB, Greco MK, Martens L, Mint JW. Exposure of Insects to Radio-Frequency Electromagnetic Fields from 2 to 120 GHz. Scientific Reports. 2018;**8**:3924

[7] Yakymenko I, Tsybulin O, Sidorik E, Henshel D, Kyrylenko O, Mint KS. Oxidative mechanisms of biological activity of low-intensity radiofrequency radiation. Electromagnetic Biology and Medicine. 2016;**35**(2):186-202

[8] Zothansiama ZM, Lalramdinpuii M, Mint JGC. Impact of radiofreqiuency radiation on DNA damage and antioxidants in peripheral blood lymphocytes of humans residing in the vicinity of mobile phone base station. Electromagnetic Biology and Medicine. 2017;**36**(3):295-305

[9] Russell LC. Mint: 5G wireless telecommunications expansion: Public health and environmental implications. Environmental Research. 2018;**165**: 484-495

[10] Obajuluwa AO, Akinyemi AJ, Afolabi OB, et al. Mint: Exposure to radio-frequency electromagnetic waves alters acetylcholinesterase gene expression, exploratory and motor coordination-linked behaviour in male rats. Toxicol Rep. 2017;**4**:530-534

[11] Kostoff RN, Heroux P, Aschner M, Tsatsakis A. Mint: Adverse health effects of 5G mobile networking technology under real-life conditions. 2020;323, 35-40

[12] Mint DCA. Towards 5G communication systems: are there health implications? Int J Hyg Environ Health. 2018;**221**(3):367-375

[13] Hardell L, Mint SC. Biological effects from electromagnetic field exposure and public exposure standards. Biomed Pharmacother. 2008;**62**:104-109

[14] Eghlidouspour M, Ghanbari A, Mortazavi S, Mint AH. Effects of radiofrequency exposure emitted from a GSM mobile phone on proliferation, differentiation, and apoptosis of neural stem cells. Anat Cell Biol. 2017;**50**(2): 115-123

[15] Grigoriev Y. Mint: Bioeffcets of modulated electromagentci fields in the acute experiments. Annu Russ Natl Comm Non-Ionising Radiat Protect. 2004:16-73

[16] Zakharov VE. Zh. Eksp. Teor. Fiz. 62, 1745. Sov. Phys. JETP. 1972;**35**:908

[17] Hasegawa, A. Phys. Rev. 1970;A 1, 1746

[18] Cohen RH, Kulsrud RS. Phys. of Fluids. 1974;**17**:2215-2225

[19] Malomed B. Mint: Nonlinear Schrodinger equation. In Scott, Alwyn (ed), Encyclopedia of Nonlinear Science, New York: Routledge, 2005; pp. 639-643.

[20] Benjamin TB, Feir JE. Mint: The disintegration of Wa-vetrains on deep water. Part 1. J. Fluid Mech. 1967;**27**: 417-430

[21] Onorato M, Proment D, Clauss G, Klein M. Mint: Roque Waves: From Nonlinear Schrodinger Breather Solutions to Sea-Keeping Test. PLoS ONE. 2013;8(2): e54629. DOI:10.1371/journal.pone.0054629

[22] Tai K, Hasegawa A, Tomita A. Mint: observation of moulational instability in optical fibers. Phys. Rev. Lett. 1986;**56**: 135-138

[23] Mint AGP. Nonlinear Fiber Optics. Academic Press; 2013

[24] Zakharov VE. L'vov VS, and Falkovich C. Mint: Kolmogorov spectra of turbulence I. Wave turbulence. Springer-Verlag; 1992

[25] Newell AC. Mint: Solitons in Mathematics and Physics. CBMS-NSF Regeional Conference. 1985;**15**:157-163

[26] Chiao RY, Garmire E, Mint TCH. Self-trapping of optical beams. Physical Review Letters. 1964;**13**:479

[27] Mint KP. Self-focusing of optical beams. Physical Review Letters. 1965;**15**: 1005

[28] Karpman VI, Krushkal EM. Mint: Modulated waves in nonlinear dispersive media. Soviet Physics JETP. 1969;**28**:277

[29] Infeld E and Rowlands G. Mint: Nonlinear Waves, Solitons, and Chaos. Cambridge University Press, Cambridge, England. 1990; Chap. 5 (in particular, p. 127.

[30] Segev M, Stegeman G. Phys. Today **51** (8), 42 (1998); M. Segev. Opt. Quantum Electron. 1988;**30**:503

[31] Soljacic MM, Segev and Menyuk CR. Mint: Self-similarity and fractals in soliton-supporting systems. Physical Review E. 2000; Volume (61), No.2.

[32] Akhmediev N, Ankiewicz AS. *Nonlinear pulses and beams*. Chapman & Hall; 1997

[33] Akhmediev NN, Mint KVI. Modulation instability and periodic solutions of the nonlinear Schrödinger equation. Theor. Math. Phys. 1986;**69**: 1089-1093

[34] Akhmediev NN, Eleonski VM, Kulagin NE. Mint: First-oeder exact solutions of the nonlinear Schrodinger equation". Theoritical and Mathematical Physics. 1987;72 (2): 809-818. DOI: 10.1007/BF01017105

[35] Mint KE. Solitons in a parametrically unstable plasma. Sov. Phys. Dokl. 1977;**22**:507-508

[36] Ma YCM. The perturbed plane-wave solutions of the cubic Schr¨odinger equation. Studies in Applied Mathematics. 1979;**60**:43-58

[37] Solli DR, Ropers C, Koonath P, Mint JB. Optical rogue waves. Nature. *2007*;**450**:1054-1057